Alternative High-Level Waste Treatments at the Idaho National Engineering and Environmental Laboratory

Committee on Idaho National Engineering and Environmental Laboratory (INEEL)

High-Level Waste Alternative Treatments

Board on Radioactive Waste Management

Commission on Geosciences, Environment, and Resources

National Research Council

NATIONAL ACADEMY PRESS

Washington, D.C.

NOTICE: The project that is the subject of this report was approved by the Governing Board of the National Research Council, whose members are drawn from the councils of the National Academy of Sciences, the National Academy of Engineering, and the Institute of Medicine. The members of the committee responsible for the report were chosen for their special competences and with regard for appropriate balance.

This work was sponsored by the U.S. Department of Energy, Contract Nos. DE-FC01-94EW54069 and DE-FC01-99EW5904. All opinions, findings, conclusions, and recommendations expressed herein are those of the authors and do not necessarily reflect the views of the Department of Energy.

International Standard Book Number 0-309-06628-X

Cover: *Above*: A 1981 photo of the calcination vessel at the New Waste Calcining Facility. The 5-foot-diameter middle section of the vessel contains the fluidized bed that is supplied with air, kerosene fuel, and radioactive waste feed. Removable nozzles, three for fuel and three for feed, are located in this middle section immediately above the metal tracks that protrude from the exterior wall and that are tapered on their bottom edge. Smaller visible piping supports instrumentation. The insulated fluidizing air supply is the large vertical piping on the left-hand side that bends to enter the bottom of the vessel. Not visible is a cyclone separator located above the calciner to remove fine particles from the gaseous effluent. SOURCE: INEEL Photograph #81-3072.

Below: An engineering artist's view of the calciner operation, showing the fuel and feed nozzles and the piping to supply fluidizing air and to remove solid calcine product. A distributor plate exists between the bottom conical section and the middle section to evenly allocate fluidizing air to the bed material (often dolomite) in the middle section. The expanded upper section allows product particles of large diameter to separate from the off-gas. SOURCE: INEEL graphic.

Additional copies of this report are available from:

National Academy Press
2101 Constitution Ave., NW
Box 285
Washington, DC 20055
800-624-6242
202-334-3313 (in the Washington Metropolitan Area)
http://www.nap.edu

Copyright 1999 by the National Academy of Sciences. All rights reserved.

Printed in the United States of America

THE NATIONAL ACADEMIES

National Academy of Sciences
National Academy of Engineering
Institute of Medicine
National Research Council

The **National Academy of Sciences** is a private, nonprofit, self-perpetuating society of distinguished scholars engaged in scientific and engineering research, dedicated to the furtherance of science and technology and to their use for the general welfare. Upon the authority of the charter granted to it by the Congress in 1863, the Academy has a mandate that requires it to advise the federal government on scientific and technical matters. Dr. Bruce M. Alberts is president of the National Academy of Sciences.

The **National Academy of Engineering** was established in 1964, under the charter of the National Academy of Sciences, as a parallel organization of outstanding engineers. It is autonomous in its administration and in the selection of its members, sharing with the National Academy of Sciences the responsibility for advising the federal government. The National Academy of Engineering also sponsors engineering programs aimed at meeting national needs, encourages education and research, and recognizes the superior achievements of engineers. Dr. William A. Wulf is president of the National Academy of Engineering.

The **Institute of Medicine** was established in 1970 by the National Academy of Sciences to secure the services of eminent members of appropriate professions in the examination of policy matters pertaining to the health of the public. The Institute acts under the responsibility given to the National Academy of Sciences by its congressional charter to be an adviser to the federal government and, upon its own initiative, to identify issues of medical care, research, and education. Dr. Kenneth I. Shine is president of the Institute of Medicine.

The **National Research Council** was organized by the National Academy of Sciences in 1916 to associate the broad community of science and technology with the Academy's purposes of furthering knowledge and advising the federal government. Functioning in accordance with general policies determined by the Academy, the Council has become the principal operating agency of both the National Academy of Sciences and the National Academy of Engineering in providing services to the government, the public, and the scientific and engineering communities. The Council is administered jointly by both Academies and the Institute of Medicine. Dr. Bruce M. Alberts and Dr. William A. Wulf are chairman and vice chairman, respectively, of the National Research Council.

COMMITTEE ON IDAHO NATIONAL ENGINEERING AND ENVIRONMENTAL LABORATORY (INEEL) HIGH-LEVEL WASTE ALTERNATIVE TREATMENTS

ROBERT C. FORNEY, *Chair*, E.I. DuPont (retired), Unionville, Pennsylvania
EDWARD AITKEN, General Electric Co. (retired), La Quinta, California
ROBERT BERTUCIO, SCIENTECH, Inc., Kent, Washington
DAVID O. CAMPBELL, Oak Ridge National Laboratory (retired), Oak Ridge, Tennessee
MELVIN S. COOPS, Lawrence Livermore National Laboratory (retired), Santa Rosa, California
DELBERT E. DAY, University of Missouri, Rolla
P. GARY ELLER, Los Alamos National Laboratory, Los Alamos, New Mexico
RODNEY C. EWING, University of Michigan, Ann Arbor
JOHN M. KERR, Innovative Technologies International, Inc., Lynchburg, Virginia
JEAN'NE M. SHREEVE, University of Idaho, Moscow
MINORU TOMOZAWA, Rensselaer Polytechnic Institute, Troy, New York

Staff

THOMAS KIESS, Study Director
SUSAN B. MOCKLER, Research Associate
LATRICIA C. BAILEY, Senior Project Assistant

BOARD ON RADIOACTIVE WASTE MANAGEMENT

MICHAEL C. KAVANAUGH, *Chair,* Malcolm Pirnie, Inc., Oakland, California
JOHN F. AHEARNE, *Vice Chair,* Sigma Xi, The Scientific Research Society, and Duke
　　University, Research Triangle Park and Durham, North Carolina
ROBERT J. BUDNITZ, Future Resources Associates, Inc., Berkeley, California
MARY R. ENGLISH, University of Tennessee, Knoxville
DARLEANE C. HOFFMAN, Lawrence Berkeley National Laboratory, Berkeley, California
JAMES H. JOHNSON, Jr., Howard University, Washington, D.C.
ROGER E. KASPERSON, Clark University, Worcester, Massachusetts
JAMES O. LECKIE, Stanford University, Stanford, California
JANE C.S. LONG, University of Nevada, Reno
CHARLES McCOMBIE, International Consultant, Gipf-Oberfrick, Switzerland
WILLIAM A. MILLS, Oak Ridge Associated Universities (retired), Olney, Maryland
D. WARNER NORTH, NorthWorks, Inc., Mountain View, California
MARTIN J. STEINDLER, Argonne National Laboratory (retired), Argonne, Illinois
JOHN J. TAYLOR, Electric Power Research Institute, Palo Alto, California
MARY LOU ZOBACK, U.S. Geological Survey, Menlo Park, California

NRC Staff

KEVIN D. CROWLEY, Director
ROBERT S. ANDREWS, Senior Staff Officer
THOMAS E. KIESS, Senior Staff Officer
JOHN R. WILEY, Senior Staff Officer
SUSAN B. MOCKLER, Research Associate
TONI GREENLEAF, Administrative Associate
LATRICIA C. BAILEY, Senior Project Assistant
MATTHEW BAXTER-PARROTT, Project Assistant
LAURA D. LLANOS, Project Assistant
PATRICIA A. JONES, Senior Project Assistant
ANGELA R. TAYLOR, Senior Project Assistant

COMMISSION ON GEOSCIENCES, ENVIRONMENT, AND RESOURCES

GEORGE M. HORNBERGER (Chair), University of Virginia, Charlottesville
RICHARD A. CONWAY, Union Carbide Corporation (Retired), S. Charleston, West Virginia
THOMAS E. GRAEDEL, Yale University, New Haven, Connecticut
THOMAS J. GRAFF, Environmental Defense Fund, Oakland, California
EUGENIA KALNAY, University of Maryland, College Park
DEBRA KNOPMAN, Progressive Policy Institute, Washington, DC
KAI N. LEE, Williams College, Williamstown, Massachusetts
RICHARD A. MESERVE, Covington & Burling, Washington, DC
BRAD MOONEY, J. Brad Mooney Associates, Ltd., Arlington, Virginia
HUGH C. MORRIS, El Dorado Gold Corporation, Vancouver, British Columbia
H. RONALD PULLIAM, University of Georgia, Athens
MILTON RUSSELL, University of Tennessee, Knoxville
THOMAS C. SCHELLING, University of Maryland, College Park
ANDREW R. SOLOW, Woods Hole Oceanographic Institution, Woods Hole, Massachusetts
VICTORIA J. TSCHINKEL, Landers and Parsons, Tallahassee, Florida
E-AN ZEN, University of Maryland, College Park
MARY LOU ZOBACK, U.S. Geological Survey, Menlo Park, California

Staff

ROBERT M. HAMILTON, Executive Director
GREGORY H. SYMMES, Associate Executive Director
JEANETTE SPOON, Administrative and Financial Officer
DAVID FEARY, Scientific Reports Officer
SANDI FITZPATRICK, Administrative Associate
MARQUITA SMITH, Administrative Assistant/Technology Analyst

Executive Summary

From 1953 until 1992, the facility near Idaho Falls, Idaho, now known as the Idaho National Engineering and Environmental Laboratory (INEEL), reprocessed a variety of nuclear fuels primarily for the recovery of uranium-235. The liquid waste from these activities, a high-level waste (HLW), was stored in subsurface stainless steel tanks, each enclosed by a concrete vault. To conserve volume, these wastes were calcined[1] starting in the early 1960's and the calcine sent to storage in partially buried stainless steel bins grouped in sets, with each set inside a separate concrete vault. The process of calcining the HLW was completed early in 1998. As the tanks were emptied of HLW, they have been used to store liquid waste from various cleanup activities at the INEEL facility. This liquid is a mixed transuranic (TRU) waste high in sodium and is referred to henceforth as sodium-bearing waste (SBW). Some of the SBW has been calcined and stored in the same bins as the HLW calcine, thereby being mixed with and converted to a HLW. For several decades, research and development activities at INEEL have studied technical alternatives for the future remediation, storage, and ultimate disposition of HLW calcine and SBW.

Several such technical alternatives are under current consideration by the U.S. Department of Energy (DOE) for the selection of one as the basis for future waste management operations. This report is the result of a National Research Council study, made at the request of DOE, to assess independently these technical alternatives. The Preface that follows summarizes the scope of this study, the committee appointed to conduct it, and the information-gathering activities undertaken. The large volume of reference materials, supplied primarily by DOE, is listed in Appendix A.

In technical options under consideration by DOE, the 4,000 m^3 of calcine would be retrieved from the bins and either

- immobilized directly into glass, cement, or glass/ceramic waste forms; or
- dissolved and treated first to separate one or more of the most radioactive components (e.g., cesium (Cs), strontium (Sr), and/or TRU radioisotopes) into a relatively small volume. Both this high-activity waste (HAW) fraction and the larger volume of lower activity waste would then be immobilized into final waste forms.

---

[1] This calcination process injected waste into a fluidized bed at elevated temperatures to evaporate the water and decompose other material constituents into "calcine," a granular ceramic.

The 5,000 m³ (as of August 1998) of SBW would be either

- calcined and added to the existing bins of HLW calcine, or
- removed from the tanks and immobilized separately, possibly after a separation step to isolate a HAW fraction.

All waste forms produced in these operations would go to storage after immobilization.

Incomplete Characterization Data

The committee found that the current chemical, physical, and radiological characterization data for HLW calcine and SBW are too incomplete and/or inaccurate for any of the types of stored waste to permit a logical selection, at present, from among the several treatment alternatives[2] available. There appears to be no realistic sampling or characterization plan at this time to improve the database. Representative sampling and analysis of actual aged calcine and SBW in current storage must be done to establish definitive bounds on critical properties, which are needed in order to design a suitable strategy and methodology for efficient separations, immobilization, and bin/tank closure operations. Important constituents include the major radioisotopes (i.e., Cs, Sr, and TRU), hazardous materials (e.g., lead, cadmium, and mercury) regulated under the Resource Conservation and Recovery Act (RCRA), and other chemical compounds that could potentially interfere with separations processes or that could provide limitations (e.g., on operational requirements, waste form durability, or waste loading) to immobilization methods.

Additional Testing Needs

For processing of either SBW or calcine (i.e., dissolution of solids and subsequent separations (e.g., Cs, Sr, and/or TRU) and/or immobilization steps), extensive testing of individual process steps is required on actual waste samples to reduce technical uncertainties associated with crucial processing parameters. However, even if satisfactory operation of each individual step could be attained under carefully controlled conditions, the objective of full-scale operation of an integrated process (i.e., containing dissolution and subsequent separations and/or immobilization steps) under realistic plant conditions will be much more difficult to meet without encountering complex operational problems, exorbitant costs, and generation of excessive amounts of secondary waste. Testing of such a fully integrated process would be required, and on a scale sufficiently large such that further scale-up challenges do not affect the likely success of the operation.

Establishment of Definitive Waste Form Specifications

To plan for such testing of an integrated operation, the separations and immobilization requirements should logically be determined by deriving them from disposal specifications for the final waste form(s). However, these disposal specifications are not yet firmly established. Where the ultimate waste disposal is to take place, the form in which waste can be received and stored in that location, and the permitting for its transportation to that location are such significant considerations affecting the waste form specifications and processing requirements that they would have to be in place before a logical design of a fully integrated process could be selected and a large-scale test made.

---

[2] A complete treatment alternative would use an immobilization step and might also use one or more chemical and physical separations processes.

Disposal Uncertainties

Many uncertainties due to currently unresolved issues of a regulatory, other legal, technical, and/or policy nature act as barriers that stand in the way of a firm designation of an ultimate disposal site and of an acceptable waste form for such a site. It is not certain at present that INEEL waste products can be disposed of in most of the disposal locations outlined in Chapter 9 because of uncertainties in (1) whether these repositories will be available for INEEL waste products and/or (2) whether the INEEL wastes are properly qualified to meet the waste acceptance criteria. Insofar as the regulatory, other legal, technical, and DOE policy requirements for the first geological repository for HLW (i.e., the planned Yucca Mountain facility) have evolved in the past two decades, it is conceivable, if not likely, that these requirements might continue to change. The requirements associated with a second HLW repository have not been established, and may differ from those of the first repository, if indeed a second repository is mandated. Only the low-level waste (LLW) disposal sites, and the Waste Isolation Pilot Plant (WIPP) for TRU LLW, are operational at present. The only offsite disposal option immediately available (i.e., without the need for a regulatory petition or special ruling) for even low-level INEEL wastes is DOE's Hanford disposal site (for non-mixed LLW material).

Selection of a Treatment Option

With the ultimate disposition pathway unknown, uncertainties from unresolved issues impeding its determination, inadequate characterization data available for both HLW calcine and SBW, and insufficient testing done to clearly reject or point to the likely success of any separations and/or immobilization process, the committee concluded that, in short, not enough information is available to make a sensible choice among technical alternatives. The ultimate disposal site(s), acceptable waste form(s), and approved transportation route(s) must be in place as part of an overall waste management strategy before any decision is made among treatment alternatives for the HLW calcine. Without these boundary conditions identified, one cannot be even reasonably certain that any option selected in the near term will both work within the present schedule (and within presently known technical risks), and also provide a waste form that will meet disposal criteria.

Recommendation on HLW Calcine

The committee could identify no significant present hazard to public health or to the environment due to the storage of solid calcine in the bins at INEEL, which have been designed to be secure for at least 500 years. The need for immediate action and a rush to select a long-term treatment option appear unwarranted, in the committee's view, especially in comparison to significant inventories of HLW at other DOE sites that are in liquid form in underground tanks, some of which have leaked.

Therefore, and since the requirements that the final waste form(s) must meet are unknown, *it is the committee's firm view that the interim storage of calcine in the bins should be maintained at least until such time as it becomes clear (1) where the material can be sent, (2) what disposal form(s) is/are acceptable, and (3) that an approved transportation pathway to the disposal site is available.* The committee supports aggressive efforts to determine the site(s), form(s), and route(s). In the meantime, the decay of the radioactive constituents of the HLW calcine would potentially reduce some processing requirements (e.g., Cs and Sr separations specifications) and minimize the exposure risks and attendant worker safety hazards, thereby reducing remediation risks and costs. Only after an approved destination for the HLW calcine is available and the requisite specifications on the waste form are known should a decision then be made to select among technical options, including that of continued interim bin storage, for calcine treatment.

If an assessment of the integrity of the bins confirms that they will be secure for a period of time comparable to their design life, then bin storage is a low-risk configuration for the near future. This course of action should be subject to continued review and updating of comparative risks as additional information may be developed in the decades to come. This recommendation does not challenge the general strategy of geologic disposal for HLW, which is an issue outside the scope of this study, but emphasizes that decisions on the ultimate fate of the INEEL HLW should be postponed pending the resolution of waste management issues noted above and pending the results of adequate risk analyses.[3]

In this recommendation to defer processing of HLW calcine until site(s), route(s), and waste form specifications are firmly established, *no time period is specified for the duration of interim bin storage*. Limitations to this time period could come from

(1) a technical assessment of bin integrity over time, and/or
(2) regulatory and other requirements.

To expand on (1), the information provided to the committee does not specify the failure mode (e.g., corrosion or seismic stability) that is the most limiting for bin integrity, and does not indicate whether 500 years signifies a mean time to failure or another design criterion. The committee recommends that during any period of interim bin storage, continuing verification of bin integrity is essential. To expand on (2), if the Licensing Requirements for the Independent Storage of Spent Nuclear Fuel and High-Level Radioactive Waste (10 CFR 72) were to apply to INEEL bin storage, then a regulatory license could be granted for up to 20-40 years, with any renewals for time periods beyond that contingent upon sufficient technical justification to satisfy the requirements for a license extension. Such matters of regulatory strategy were not examined in detail by the committee, and would require attention in the event that resolution of ultimate disposal site(s), route(s), and waste form specifications is not attained in the near future.

Recommendation on SBW

The SBW is stored as a liquid in tanks that do not meet regulatory approval standards for long-term storage. In the committee's view, the DOE should solidify the SBW as soon as practicable. To meet this objective, solidification options other than calcination (which results in the counterproductive conversion of SBW to HLW when SBW calcine is mixed in storage with HLW calcine) should be identified and one of them selected. This effort should consider processes developed for similar DOE wastes at other sites that could be adapted to the SBW. The remediation goal should be to produce a solid at relatively low cost and personnel risk that is suitable for shipment to the now-operating Waste Isolation Pilot Plant (WIPP) or to another TRU waste site if one becomes available. Several solidification options for SBW are offered in Chapter 12 of this report. Drawing upon the backgrounds of personnel available at other DOE sites would be appropriate in considering these and other possible approaches.

---

[3] If bin storage of HLW calcine persists for more than a generation, then this recommendation does counter the general view (OECD, 1995) of having the present generation "dispose" of its own long-lived radioactive waste in geologic repositories; instead, for INEEL HLW calcine, the present generation would "manage" this waste. Such a bin storage strategy is consistent with the "stepwise implementation of plans for geological disposal" that might take place "over several decades" and is also consistent with the admission of "the possibility that other options could be developed at a later stage" that is expressly stated in OECD (1995: p. 9).

Closure Specifications

Residuals of SBW and HLW calcine will remain at the INEEL site regardless of any option considered to process these wastes. The amount of residual waste left in place should be assessed in various scenarios to determine relevant "trade-offs" such as the costs and benefits of risk reductions that are achieved. These assessments of closure options should be used to define tank closure criteria and specifications.

Risk Analysis Perspective

As a final observation, the committee believes that, along with good science and engineering, a major consideration in deciding how (and *whether*) to process any radioactive waste for long-term conditioning is that of the risks being added and/or mitigated. The fundamental purpose of environmental regulations (such as those of RCRA, the Comprehensive Environmental Response, Compensation, and Liability Act, and United States Nuclear Regulatory Commission directives) and radioactive waste policy legislation must be, in the committee's view, minimization of risk to human health and the environment in a cost-effective and meaningful manner. A driving consideration in deciding upon a radioactive waste management strategy should be an identification, definition, and evaluation of the "trade-offs" (i.e., comparative risks) for the alternatives being considered, including those of limited or no processing. Such risk assessment calculations provide information on risk reduction strategies and are required to decide among alternatives in an informed and objective manner. Consequently, the committee believes that a risk analysis for the actions recommended above for both HLW calcine and SBW should be conducted promptly, and should include a comparison of the risks associated with INEEL HLW calcine and SBW to the risks associated with site inventories of other radioactive wastes. A sufficiently rigorous analysis should be performed to establish the current risks and to assess the changes in risk due to treatment options. In the committee's view, at least until the issues identified here are resolved, the risks (and costs) of repackaging the HLW calcine, with no certainty that it can ever be shipped outside of Idaho, may far exceed the risks (and costs) of continued interim bin storage.

Preface

This report is the product of a National Research Council (NRC) committee study initiated at the request of the U.S. Department of Energy (DOE) to examine technical options for treating radioactive high-level waste (HLW) at the Idaho National Engineering and Environmental Laboratory (INEEL). This HLW was produced in the reprocessing of approximately 44 metric tons of spent nuclear fuel during 1953-1992. This study was done to provide information for a future DOE decision on which of several possible waste management approaches should be adopted in site plans and activities to convert the HLW into forms suitable for transport and/or disposal.

This committee study was organized within the NRC's Board on Radioactive Waste Management (BRWM) and was conducted by an 11-member committee. Committee members were chosen for their expertise in relevant technical disciplines such as nuclear and fluoride chemistry, process engineering, nuclear materials handling, waste form development, and risk assessment. As is normal Academy practice, committee members did not represent views of their institutions, but formed an independent body to author this report.

To conduct the study, the committee gathered information in selected ways, principally through meetings and literature review. The committee met twice in Idaho Falls, Idaho, during August and October of 1998, to hear from DOE and its contractors, invited guests, representatives of the Idaho regulatory oversight authority, and members of the public. At the invitation of the committee, seven "technical experts" attended the sessions of the first meeting that were open to the public and prepared trip reports of their impressions, which appear as Appendix B in this report. The committee also reviewed the DOE literature, listed in Appendix A. Committee members prepared this report using these inputs and their collective knowledge and experience.

The remainder of committee meetings and interactions were devoted to the preparation of the report, representing the consensus view of the committee and responsive to the following Statement of Task:

The purpose of this project is to perform an independent assessment of the INEEL High-Level Waste Program. This assessment will include the following: (1) examination of the set of treatment options chosen by the DOE and identification of other alternatives that DOE might consider; (2) technical analysis of the assumptions, criteria, and methodology used in selecting the baseline waste treatment option and any additional action that would improve its technical validity; (3) evaluation of environmental and technological risks associated with the chosen set of treatment options; and (4) feasibility of the treatment options given the regulatory framework and compliance agreements. A final report would provide technical assessments of alternatives to help the DOE select a preferred path forward and meet their regulatory milestones on schedule.

Acknowledgements

Thanks are due to the seven expert consultants noted below who attended the first information-gathering meeting during August 17-19, 1998, and provided written trip reports for the committee's consideration. Their reports are in Appendix B, and their brief biographies in Appendix D.

Mr. John Roecker, consultant, Tank Waste Remediation System at Hanford;
Dr. Anne Smith, vice president, Charles River Associates;
Dr. David Clark, Professor of Materials Science and Engineering, University of Florida;
Dr. Edward Lahoda, Advisory Engineer, Westinghouse Science and Technology Center;
Dr. K. K. Sivisankara Pillay, senior staff member, Los Alamos National Laboratory;
Mr. Ernest Ruppe, (retired) vice president of petrochemicals, E.I. DuPont; and
Dr. Barry Scheetz, Professor of Materials and Senior Scientist, The Pennsylvania State University.

The views of these individuals, as contained in their trip reports, are not to be construed as those of the organizations with whom they have been or are affiliated.

Thanks are also due to Dr. Martin Steindler and Mr. John Taylor of the Board on Radioactive Waste Management (BRWM) for their roles as liaisons between the BRWM and the committee during the study.

This report has been reviewed by individuals chosen for their diverse perspectives and technical expertise, in accordance with procedures approved by the National Research Council (NRC) Report Review Committee. The purpose of this independent review is to provide candid and critical comments that will assist the NRC in making the published report as sound as possible and to ensure that the report meets institutional standards for objectivity, evidence, and responsiveness to the study charge. The content of the review comments and draft manuscript remains confidential to protect the integrity of the deliberative process. Thanks are due to the following individuals, who are neither officials nor employees of the NRC, for their participation in the review of this report:

Richard Conway, Union Carbide Corporation, retired
Jurgen Exner, JHE Technology Systems, Inc.
Peter Hayward, Atomic Energy of Canada Limited, retired
Darleane Hoffman, Lawrence Berkeley National Laboratory
Lee Hyder, Savannah River Technology Center, retired
Antoine Jouan, Commisariat à l'Energie Atomique
Milton Levenson, Bechtel International, Inc., retired
John Mackenzie, University of California
D. Warner North, NorthWorks, Inc.
Frank Parker, Vanderbilt University

While the individuals listed above have provided many constructive comments and suggestions, it must be emphasized that responsibility for the final content of this report rests entirely with the authoring committee and the NRC.

Contents

Executive Summary, vii

Preface, xiii

1   Introduction, 1
2   Calcine Characterization, Retrieval, and Dissolution, 15
3   Physical and Chemical Separations, 29
4   Treatment Options for Sodium-Bearing Liquid Waste, 43
5   Vitrification, 47
6   Cementation, 55
7   Other Waste Forms, 61
8   Tank and Bin Closure, 71
9   Constraints Imposed by Disposal Options, Regulations, and Cost, 81
10   Inconsistencies Associated with Current Plans, 91
11   What Should Be Done: INEEL HLW Calcine, 97
12   What Should Be Done: Sodium-Bearing Liquid Waste, 101
13   Summary of Conclusions and Recommendations, 107

References, 121

Appendices
A   Documents Received During This Study, 129
B   Trip Reports of Technical Experts, 141
C   Biographical Sketches of Committee Members, 167
D   Biographical Sketches of Technical Experts, 171
E   Glossary, 175
F   Acronyms, 179
G   Portions of the 1995 Settlement Agreement, 181

1

Introduction

The problem forming the subject of this report is the long-term disposition of radioactive high-level waste (HLW[1]) at the Idaho National Engineering and Environmental Laboratory (INEEL). This report assesses alternative ways to convert this waste into forms that are suitable for interim storage and eventual transport to and disposal in geologic repositories or other appropriate facilities. This chapter summarizes the past activities at INEEL that generated this waste, current waste inventories, current waste management practices and facilities, and future plans. This background material defines the starting point for all treatment options, which are comprised of the technical process steps considered in greater detail in the chapters that follow.

WASTE GENERATED FROM PAST REPROCESSING PRACTICES

INEEL has, among its other missions, reprocessed a variety of nuclear fuels, primarily for recovery of the fissile isotope uranium-235 (^{235}U). These fuels generally used highly enriched uranium (i.e., a blend of ^{235}U and uranium-238 with the isotopic abundance of ^{235}U in excess of 20 percent by weight) and a variety of cladding materials (aluminum, zirconium, stainless steel, and graphite). In reprocessing, these fuels and their cladding were dissolved in acidic solutions (nitric acid, hydrofluoric acid, and sulfuric acid were used). This permitted species of interest (e.g., ^{235}U, as well as, from time to time, neptunium-237, krypton-85, barium-140, and fission product xenon) to be separated and recovered (Knecht et al., 1997).

The liquid acidic wastes from these activities were stored in subsurface tanks constructed of stainless steel and surrounded by concrete containers called vaults. In time, this liquid waste was converted to a solid form in calciners and stored in bins on site. Various calcination facilities were developed and tested in the 1950s, culminating in the Waste Calcining Facility (WCF), a fluidized bed unit constructed during 1958-1961 and operational on radioactive feed during 1963-1981 (Knecht et al., 1997). The New Waste Calciner Facility (NWCF), another fluidized bed unit with increased capacity and reduced maintenance requirements, began operation in 1982.

Both the liquid reprocessing waste and the solids resulting from its calcination are classified as HLW, as that term is defined by both the U.S. Department of Energy (DOE) and the U.S. Nuclear Regulatory Commission (USNRC), because they come from the "first-cycle

---

[1] A complete list of acronyms used in this report appears in Appendix F.

raffinate" produced in nuclear fuel reprocessing.[2] This raffinate contains radioactive fission products and actinides.

From 1953 to 1992, these reprocessing activities and associated waste management practices were done at the INEEL's Idaho Chemical Processing Plant (ICPP).[3] Reprocessing activities ceased in 1992. Further details of reprocessing campaigns, fuel types, equipment upgrades, and chemical additives are found in Knecht et al. (1997). The INEEL HLW practices differed from those at other DOE sites in two important ways:

1. The acidic raffinates were not neutralized with sodium hydroxide, which would have increased their volume and possibly precluded their calcination.
2. Instead, these raffinates, with added chemicals (such as calcium, whose addition neutralized excess acid and precipitated fluoride as CaF_2), were calcined directly to a solid form (Knecht et al., 1997).

SOLID WASTE IN BINS

As noted above, the calciners operated to generate solid calcine that is piped directly into bins. The solid waste bins are partially buried stainless steel bins grouped inside concrete vaults (see Figure 1.1), and they are designed to last 500 years (Palmer, 1998; Dirk, 1994[4]; Schindler, 1974).[5] Three to seven bins built adjacent to each other and within the same vault form a "bin set," and the piping from the calciner distributes calcine into each bin of a "bin set" (see Figure 1.2). As of August 1998, five of the seven bin sets are filled, one is partially full, and one is empty. The 1998 inventory of calcine was approximately 4,000 m^3 (Palmer, 1998).

Calcines have composition dependent upon the nuclear fuel, cladding materials, and chemicals added in reprocessing and calcination. Alumina calcines, comprising approximately 800 m^3 of the 1995 inventory (Brewer et al., 1995), are predominantly alumina (Al_2O_3) from dissolution of aluminum-clad spent fuel. Zirconia-based calcines, comprising approximately 3,000 m^3 of the 1995 inventory (Brewer et al., 1995), contain zirconia (ZrO_2) from the reprocessing of zirconium-based fuels (Knecht et al., 1997) and other constituents such as calcium fluoride (CaF_2) and alumina (Al_2O_3). The fluoride content of zirconia-based calcines comes from the hydrofluoric acid that was used to dissolve the fuel, and the aluminum and calcium content comes from the addition of compounds containing these elements in order to complex the fluoride. Fission products are less than 1 percent by weight but account for most of the radioactive curie content (Garcia, 1997). In addition to radioactive species, hazardous chemical constituents (e.g., compounds containing mercury), added during reprocessing, waste management, and calcination operations, may be present in the calcine. Further details on calcine characterization are found in Garcia (1997).

---

[2] High-level waste is (1) irradiated reactor fuel; (2) liquid wastes resulting from the operation of the first-cycle solvent extraction system, or equivalent, and the concentrated wastes from subsequent extraction cycles, or equivalent, in a facility for reprocessing irradiated reactor fuel; and (3) solids into which such liquid wastes have been converted (USNRC 10 CFR 60 and proposed 10 CFR 63).

[3] In May 1998, the ICPP was renamed the Idaho Nuclear Technology and Engineering Center (INTEC).

[4] This reference estimates corrosion of the interior steel bin walls of only 5 one-thousandths of an inch ("mils") in 500 years.

[5] Five hundred years is also used as a design value in Dahlmeir et al. (1998), Engineering Design File EDF-BSC-007.

FIGURE 1.1(a) Construction of Bin Set 4 of the Calcine Solid Storage Facilities at INEEL in 1976. One of the 304L stainless steel, 12-foot outer diameter, 55-foot tall bins is being placed in the concrete vault. SOURCE: INEEL photograph #76-4544.

FIGURE 1.1(b) The photograph on the left shows Bin Set 5 and Bin Set 6 of the INTEC Calcine Solid Storage Facilities. The scale-model of Bin Set 6 on the right depicts construction details and the arrangement of bins inside each of the seven bin sets. SOURCE: INEEL photograph #ICPP-S-13030.

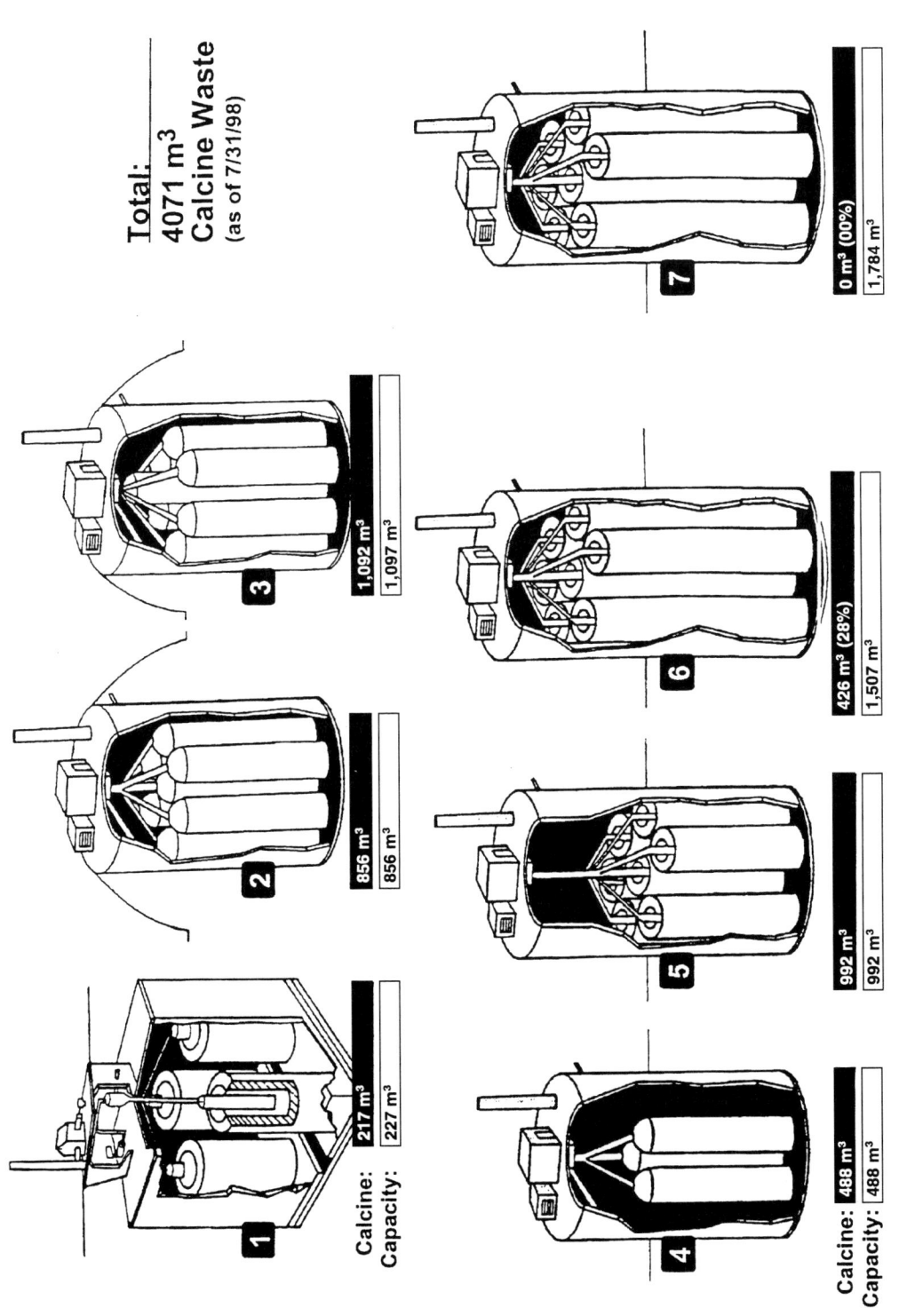

FIGURE 1.2 A diagram showing the arrangement of bins inside each of the seven bin sets. SOURCE: Palmer, 1998.

The NWCF has been shut down periodically for maintenance and recently has been targeted to operate in the future (i.e., beyond June 1, 2000) only if upgrades are performed on the emissions monitoring system, as required by the recently promulgated Maximum Achievable Control Technology (MACT) rule of the Environmental Protection Agency (EPA).[6] During past reprocessing campaigns, radioactive species (e.g., ruthenium isotopes ^{103}Ru and ^{106}Ru, which were produced in fuel as fission products and which were volatile in oxide forms) have been present in off-gas emissions (Knecht et al., 1997). Hazardous chemical effluents include mercury (Hg) compounds, carbon monoxide (CO), and NOx species, the last of which appears as an orange plume from the stack. Hg and CO emissions are regulated under MACT rules. The NOx species can interfere with the sampling and removal of Hg and organic effluents (Kimmel, 1999a).

LIQUID WASTE IN TANKS

In February 1998, the last of the liquid HLW generated from past reprocessing practices was calcined. An inventory of liquid waste remains, however, because other waste streams, such as from site decontamination and decommissioning activities have generated sodium-bearing waste (SBW). This waste is mixed transuranic waste, rather than HLW, but, because of its radioactivity and large volume, the site managers found it convenient to store this waste in the underground tanks and thus manage it as if it were HLW. As of August 1998, the amount of this liquid waste was 5,000 m^3 (1.3 million gallons), distributed in approximately half of the 11 underground tanks (Palmer, 1998). The radioactivity of the inventory of SBW in the tanks is approximately an order of magnitude less than the radioactivity of the inventory of solid HLW calcine (Olson, 1998b; DOE, 1997b: Table 2.12). Since February 1998, some SBW has been calcined using aluminum nitrate as an additive, and the resulting calcine was added to the HLW calcine in the bins. This process creates additional HLW through mixing, and the possibility that this process may be continued caused the committee to regard SBW treatment as part of the task of this study.

The underground tanks do not have secondary containment liners that meet current regulatory requirements of the Resource Conservation and Recovery Act (RCRA). Thus, because of the way they were constructed (see Figure 1.3), it cannot be assumed that the tanks will continue to hold the liquids in the event of seismic disruptions or a breach of the steel tank wall. Although detection of a leak is possible (Figure 1.4), the tanks cannot serve as a permanent RCRA-approved storage location for the liquid waste. Moreover, the actinide constituents are in sufficient concentration to enable this waste to be classified as transuranic waste, for which shallow subsurface disposal (as in the tanks) is inappropriate. A current regulatory milestone, from a "Notice of Noncompliance Consent Order," is to cease the use of all tanks by the year 2012 (Wichmann, 1998a). This would entail removal and/or solidification of all liquid waste contents.

OTHER HLW: SPENT NUCLEAR FUEL INVENTORIES

The HLW under consideration in this report is the inventory of solid calcine and liquid SBW (the latter of which is not classified as HLW but which is managed as HLW by

---

[6] The MACT rule under consideration here is that for hazardous waste incinerators regulated under the Clean Air Act Amendments of 1990. The calciner is designated as an incinerator.

FIGURE 1.3 The 1951 construction of two 300,000-gallon stainless steel liquid waste tanks, WM-180 and WM-181. Monolithic, reinforced concrete vaults poured in place in an octagonal shape surround these tanks. Four other tanks are surrounded by poured-in-place concrete vaults of square shape. Octagonal concrete vaults of "pillar and panel" construction surround the remaining 5 of the 11 steel tanks. SOURCE: INEEL photograph #51-3374.

Introduction

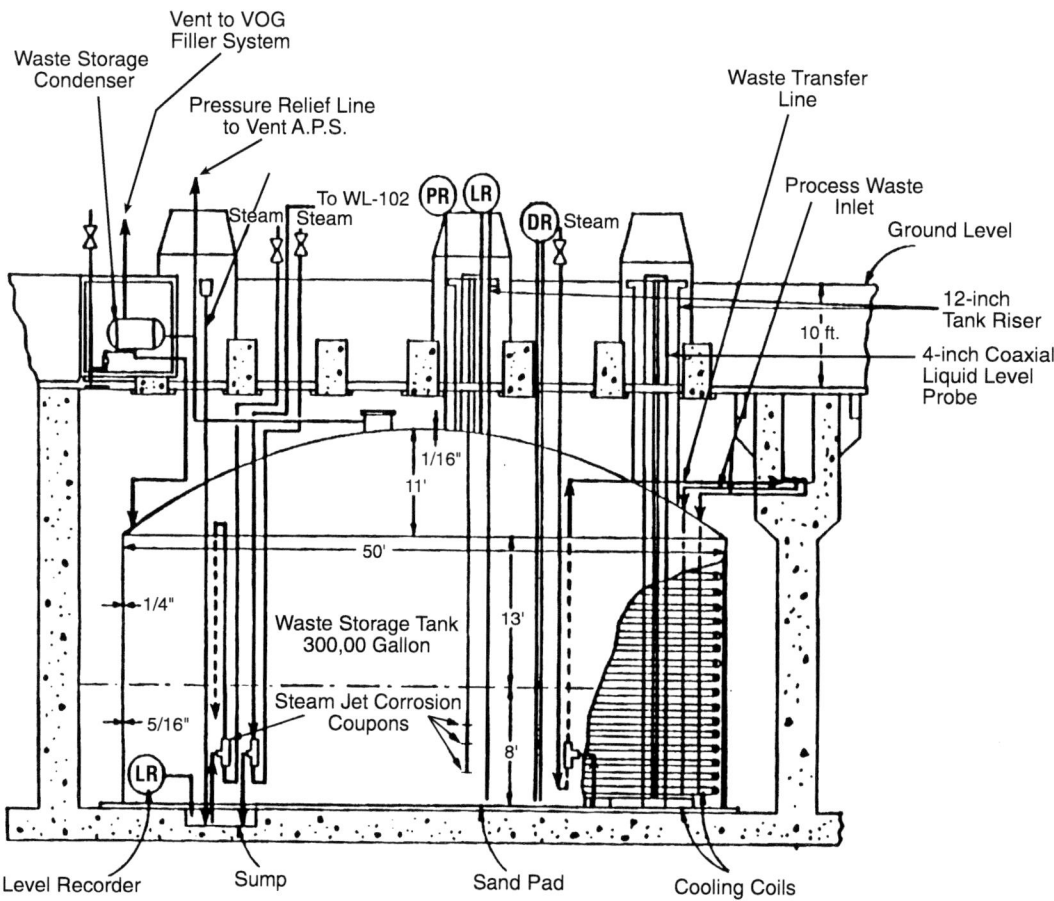

FIGURE 1.4 The design of a 300,000-gallon tank. Some of the indicated features are the leak detection capability via a sump, seismic on isolation from the concrete vault via a sand pad, corrosion coupons, and piping (i.e., cooling coils, risers, and transfer lines). SOURCE: From Palmer, 1998).

DOE, with the rationale that its characteristics are similar to the HLW liquid raffinate that was calcined in the past). Site inventories of spent nuclear fuel (SNF) that have not been reprocessed as described previously are also classified as HLW but are not included in the scope of this study. Therefore, the term HLW as used in this report will denote only the HLW calcine and SBW.

However, the management and ultimate disposition of INEEL HLW is linked to DOE disposal plans for other waste streams, principally SNF. The SNF inventories that can be compared and contrasted to the INEEL HLW calcine and SBW include (a) DOE SNF stored on the INEEL site and generated from on-site DOE reactors that operated in the past in reactor development programs and defense-related missions, (b) Navy SNF also stored on the INEEL site and generated from reactors in the Navy nuclear program, and (c) SNF planned for storage at the INEEL site and generated from other, off-site sources. As used in this report, these DOE SNF inventories are loosely labeled as "defense" in nature and are discussed only insofar as current DOE plans destine the HLW calcine and DOE SNF for co-disposal in the geologic repository under development for commercial SNF from nuclear power plants.

SNF is stored on site in wet and dry storage (specifically, in six wet pools, one dry storage pad, one dry storage vault, and one below-grade dry storage caisson) (Kimmel, 1999a). The present SNF inventory at INEEL is 232 MTHM,[7] at a total mass of 1,874 metric tons, within a volume of 558 m^3, and with an inventory of approximately 50 million (5×10^7) Ci (Kimmel, 1999a). This inventory of fuel comes from on-site DOE reactors that have operated in the past and from off-site reactors of both Naval and non-Naval origin, and continues to grow as new shipments of fuel in the latter two categories are received. The inventory of SNF from off-site non-Naval sources (i.e., other DOE sites, universities, and foreign research reactors) is estimated to grow, over the next 20 years, to total about 53 MTHM, representing a mass of 145 metric tons, a volume of 34.5 m^3, and a curie inventory[8] of 10 million (1×10^7) Ci, by the year 2020 (Kimmel, 1999a). The SNF inventory from off-site Naval sources, currently comprising 15 MTHM of the total site inventory, will grow to 65 MTHM by 2035, to a total mass of 4,400 metric tons, a total volume of 900 m^3, and a total radioactivity of approximately 300 million (3×10^8) Ci[9] (Elmer Naples, private communication with Thomas Kiess, February 1999).

These numbers indicate that the current SNF inventory is largely composed of contributions from on-site DOE and off-site Naval reactors. These SNF figures are provided here only for perspective in a comparison to similar quantities estimated for the calcine. The total HLW calcine radioactivity is approximately 40 million Ci (Garcia, 1997: corrected Table 4), with a total volume of approximately 4,000 m^3, which, using 1.4 g/cm^3 as an average density (Garcia, 1997), represents approximately 6,000 metric tons. This HLW calcine comes from the processing at INEEL of 44 metric tons of uranium at "beginning of life" (Kimmel, 1999a). Therefore, this calcine inventory is comparable to (i.e., within an order of magnitude of) the current and future INEEL SNF inventories, as measured by radioactivity, volume, or mass. Although the HLW calcine is comparable to site SNF inventories by these measures, other features, such as the content of fissile and/or long-lived radionuclides, may differ in ways that would provide a technical rationale for different management and disposition strategies.

---

[7] MTHM = metric tons of heavy metal, a unit of measure for uranium-bearing nuclear materials such as SNF.
[8] This inventory accounts for a select number of radionuclides, including the long-lived species and some of the short-lived species (^{137}Cs, ^{3}H, ^{85}Kr, and ^{90}Sr are included). The curie value listed is the projected activity of these species at the year 2030.
[9] This number is only approximate and refers to the typical fuel radioactivity approximately 5 years after shutdown.

OTHER INVENTORIES OF RADIOACTIVE WASTE

As with other large DOE sites, the INEEL also contains buried transuranic and low-level wastes. The radioactivity, volume, and mass of these wastes, and the environmental and safety risks that they pose, would provide additional perspective on relative risks than is provided here by the comparison of the INEEL HLW to the SNF inventory alone. The committee has insufficient information to devise this comparison of all site inventories of radioactive wastes, but simply notes the utility of such a comparison for informed judgment concerning which problems pose the greatest risk and which, of all possible site remediations, provide the greatest reductions in risk.

FUTURE PLANS AND CONTEXT FOR THIS STUDY

A "Settlement Agreement" reached in 1995 resulted in a court order signed by then-Governor of Idaho Philip Batt that provided milestones for DOE activities pertaining to SNF and HLW (and mixed waste, both transuranic and otherwise). To paraphrase the terms of this agreement (see Appendix G), by the year 2035 the INEEL SNF and HLW are to be in forms suitable for out-of-state transport and ultimate disposal, as a condition of the continued receipt of Naval spent fuel into Idaho. These terms are subject to change if a future Environmental Impact Statement (EIS) were to show merit in renegotiating the 2035 milestone (Lodge, 1995).

An EIS is part of the process mandated by the National Environmental Policy Act (NEPA) legislation begun in 1969 that applies to federal facilities. An EIS would examine a range of alternative strategies for addressing a remediation problem that has potential environmental impacts, and analyze these options using several criteria (e.g., public health and environmental consequences, as applied to projected operations and hypothetical accident scenarios and expressed in risk terms). The EIS is a source of information, along with other measures such as assessments of worker safety, for decisionmakers to consider in selecting among various remediation options. The DOE-EM program, created in 1989 to clean up the sites and wastes of the former nuclear weapons complex, routinely conducts EIS evaluations prior to the construction of major facilities.

The DOE-EM is developing an EIS that addresses technical alternatives for the conversion of INEEL's HLW and SBW into waste forms suitable for off-site, out-of-state transport. The EIS evaluations would provide input into a decision as to which of many treatment options to pursue for the inventories of solid HLW and liquid mixed transuranic waste, prior to the selection of one option and the subsequent commitment of resources. To meet a 2035 deadline, a decision in the near future would allow sufficient time for process development and for construction and operation of large facilities to process the current waste into different forms.

PROCESS OPTIONS

The purpose of treating the INEEL HLW would be to convert the present calcine and sodium-bearing liquid into material forms suitable for interim on-site storage and for eventual transport and disposal in repositories or other facilities. In current program plans, radioactive constituents of the HLW that have high activity and/or long half-lives are to be disposed of as high-activity waste (HAW) in a geologic repository. The less radioactive and non radioactive constituents, if separated from the HAW component and sufficiently decontaminated to meet regulatory approval as no longer necessary to manage as HLW, could be disposed of in other, shallow, land facilities, either on- or off-site. Therefore, segregation of high- and low-activity fractions is one of the process steps currently under consideration by the INEEL HLW program. A succession of such steps, if properly designed and integrated into a treatment system, would produce final waste forms with radiological and material properties tailored to meet requirements imposed by long-term disposal conditions.

For processing INEEL calcines and liquid waste into other forms, several technical options exist, as summarized briefly here. One option is to mix waste with appropriate cement-forming additives to make a cement waste form. Another option is to dissolve calcine into aqueous-based solutions; separate radioactive constituents such as ^{137}Cs, ^{90}Sr, and actinides; and convert the remaining large volume of liquid waste into a low-level grout, while solidifying the fraction containing high-level and/or long-lived radioisotopes into a smaller volume in a glass or cement waste form. A third option is to directly convert calcine into a vitrified product, without dissolution or separation of selected constituents. A fourth option is to convert calcine into an alternative waste form, such as a glass/ceramic, either directly or after a separation operation. A "No-Action" or "Minimal Action" option would involve long-term storage of calcine in the bins. Most of these options involve closure of both the tanks and bins.

These processing alternatives are being evaluated by DOE and its contractors. They are summarized in DOE (1998) and Russell et al. (1998), with further detail found in Lopez and Kimmett (1998); Dafoe and Losinski (1998); Russell and Taylor (1998); Landman and Barnes (1998); Lee and Taylor (1998); Dahlmeir et al. (1998); Spaulding (1998a,b); and Fluor Daniel (1997a,b), which build upon earlier systems analysis studies (Palmer et al., 1994; Murphy et al., 1995; Palmer et al., 1998), program plans (LMITCO, 1996), and evaluations of alternatives (Palmer, 1996). These technical alternatives are represented schematically in Figure 1.5. Further discussion in this report focuses on the individual technology components (represented by the boxes in Figure 1.5) and on the overall strategy of waste retrieval and remediation that these alternatives are designed to implement.

ORGANIZATION OF THIS REPORT

At the request of DOE, this committee study of the National Académies[10] was conducted to provide an independent review of these treatment alternatives and to consider other possible courses of action. This report is intended to assist DOE and its contractors by providing input prior to the publication of the final EIS. As noted in the Preface, which summarizes how this study was conducted to address the committee's "Statement of Task," the committee reviewed the references cited above and other literature that is listed in Appendix A. The committee also held two information-gathering meetings in Idaho Falls, Idaho, to consider the past work and current plans of DOE and its contractors. The committee invited a group of "technical experts" to attend the first of these meetings and to supply written comments in the form of trip reports that are shown in Appendix B.

Chapters 2 through 8 discuss in detail the important technical steps that would be undertaken in one or more of the treatment alternatives. These steps are as follows:

- retrieval of calcine from bins, and possible blending of calcine from different bins (Chapter 2);

---

[10] The honorific societies of the National Academy of Sciences, National Academy of Engineering, and Institute of Medicine, together with the National Research Council, are a collection of institutions that are referred to here as "the National Academies."

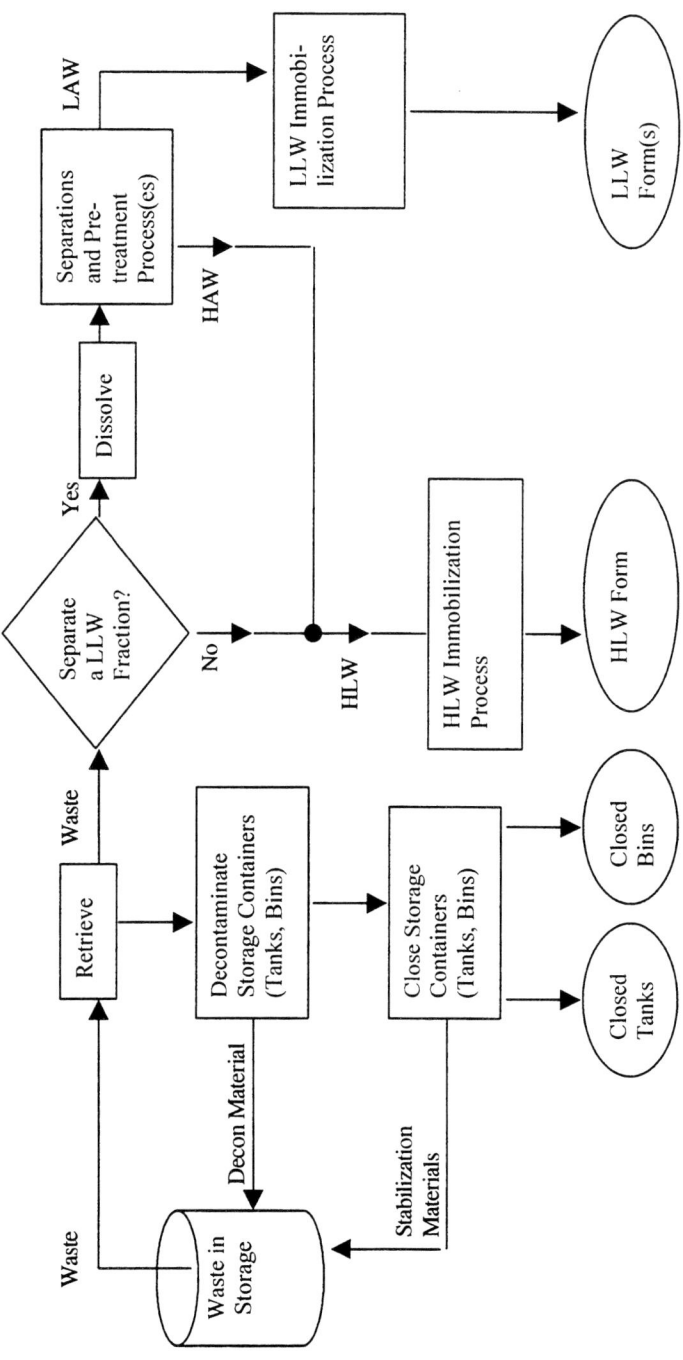

FIGURE 1.5 This diagram depicts in generic form the technical alternatives under consideration for the INEEL HLW calcine and SBW. The goal of each alternative is the conversion of initial waste products into appropriately immobilized forms. Each treatment alternative involves retrieval of waste and subsequent decontamination and closure of waste storage containers. Some alternatives use dissolution and processing steps to separate a low-level waste form. All alternatives immobilize a HLW form for onsite storage and eventual disposal.

- dissolution of calcine (Chapter 2);
- possible separations of selected radioactive species such as ^{137}Cs, ^{90}Sr, and actinides (Chapter 3);
- processing of SBW (Chapter 4);
- vitrification (Chapter 5);
- cementation (Chapter 6);
- production of glass/ceramic and other waste forms (Chapter 7); and
- closure of the tanks and bins (Chapter 8).

These process steps are evaluated in the above chapters to assess what *can* be done with current technical capability. Chapters 2 through 4 discuss ways to convert the two existing waste forms (i.e., solid calcine particles and liquid SBW) into waste streams that are ready for solidification. Chapters 5 through 7 discuss various solidification methods. Chapter 8 describes the process steps needed to remediate the unrecovered waste from the tanks and bin sets and to provide an in-situ closure of these structures.

Relevant regulations are not only those of the DOE and USNRC for radioactive HLW, but also those of the EPA (principally RCRA) that apply to the hazardous chemical constituents of INEEL HLW and SBW. Chapter 9 briefly discusses these regulatory constraints and considerations of cost, issues that influence a final decision. The full context for this decision includes multiple criteria (e.g., worker risk, public safety, and environmental impact) and multiple and potentially competing objectives (e.g., responsible waste management consistent with INEEL's future plans and missions, local interest in pristine cleanup versus national interest in reducing total program cost), as well as the technical alternatives. These issues pose challenges for program planning and management strategy, as discussed in Chapter 10, and raise the question of what *should* be done, which is discussed in Chapters 11 and 12 by way of examining the technical options favored by the committee for the calcine and the sodium-bearing waste, respectively.

Chapter 13 summarizes the most important findings, conclusions, and recommendations stated in each of the preceding chapters.

2

Calcine Characterization, Retrieval, and Dissolution

The high-level waste (HLW) calcine consists of powdered ceramic particles of various sizes and chemical compositions. The range of properties is the result of compositional differences of the inputs (namely, the reprocessing waste and chemicals added to this waste "feed" solution) to the calciner. During fluidized bed calcination, particles (usually dolomite) in the fluidized bed were coated multiple times with layers of compounds from the waste stream.[1] The high heat caused gaseous species (e.g., water and some nitrates) to evolve. The output is granular calcine particles with a range of physical, chemical, and radiological characteristics.

These properties affect the operations discussed in this chapter. Any treatment of the HLW calcine requires adequate characterization and retrieval from the bin sets where it is currently stored. Blending of various calcine types is projected to occur. Aqueous-based processing can only proceed if the calcine is dissolved in acid solution. These important characterization, retrieval, blending, and dissolution steps are discussed in this chapter.

As stated in Chapter 1, after the HLW was converted to calcine, it was transferred pneumatically to large cylindrical stainless steel bins for storage. Several such bins comprise a bin set located within a single concrete containment structure, or vault. There are six bin sets containing calcine and one that is empty. The total volume of the six bin sets in use is approximately 180,000 ft^3 (5,100 m^3), and the volume of calcine contained in them is in excess of 130,000 ft^3 (3,700 m^3)[2] (Palmer, 1996; INEEL, 1997; Lopez and Kimmett, 1998). The calcine varies substantially in composition, both between bins and in different regions within a given bin. However, most calcine composition data are deduced from calciner feed data. Some calcine characterization data are based on pilot tests with cold simulants. There are few data for actual calcine, and only two samples of stored calcine have been retrieved from bins, collected in 1979 (Staples et al., 1979; Garcia, 1997).

The primary risk from retrieval, with respect to both environmental cleanup and health effects, appears to result from the possible release of radioactive and toxic particles. Dry calcine contains particles of various sizes, some of the right size to constitute a significant respirable hazard. Therefore, appropriate measures will have to be taken to ensure that releases cannot occur during sampling, installation of equipment for retrieval, and the actual

---

[1] In long calciner runs, after approximately a week of operation, the bed particles were used up and not needed as a substrate; that is, the chemical constituents in the radioactive waste "feed" stream were a sufficient input to solidify upon heating to form calcine particles.
[2] Here and in other places English units (rather than metric) are shown when they are the units used in the referenced literature.

retrieval and subsequent handling. These operations are complicated by the fact that the bins were not specifically designed to provide for sampling or retrieval operations.

The characterization data provided to the committee were primarily from a report (Garcia, 1997) that was found by the committee to contain numerous errors (see the Adequacy of Existing Information section later in this chapter), to the point that its credibility and usefulness are questionable. However, because of the lack of alternative data, those in Garcia (1997) are used for most of the following discussion. Other data sources are specifically noted where discussed.

CALCINE CHARACTERIZATION

Calcine properties are deduced almost entirely from three sources of data: (1) samples of the solutions fed to the calciner, (2) cold (i.e., nonradioactive) calciner pilot plant testing with simulant feed, and (3) grab samples of real calcine. The majority of the work was done several years ago and does not include aging effects.

Regarding the third source mentioned above, the only samples of actual calcine that have been retrieved from bins consist of two core samples collected from the second bin set (Calcined Solids Storage Facility (CSSF) #2) in 1979. Staples et al. (1979) indicate that the two cores were full-length, with the sampler (a string of 28 sample tubes on a drill rig) in excess of 10 m long; another source (Garcia, 1997) indicates (most likely erroneously) that the cores were 1 m long. The calcination heat source was changed from indirect liquid-metal heating to in-bed combustion at about the time this waste was calcined, and no samples of calcine produced by in-bed combustion have been retrieved from bins. In-bed combustion generates both oxidizing and reducing chemical environments in different regions, which could affect calcine properties.

A calcine sample from the output of the calciner was collected in 1993 (Garcia, 1997: p. 14). This is also available for testing.

Although physical properties are reported (i.e., Table 2-1), they appear to be based on pilot tests with cold surrogates. Few data are available for long-aged calcines. Chemical compositional data (i.e., Tables 2-2 and 2-3) appear to be based largely on material from pilot-scale tests for alumina-based and zirconia-based feeds, and may not represent the full range.

The term "zirconia-type calcine" is misleading because it contains two to four times as much Al, Ca, and F as Zr.[3] Chemical data for the 1979 samples of actual calcine (Garcia, 1997: Table 5) are in limited agreement with Table 2-2, and show that there is significant variation in composition vertically within the core sample. Data collected in 1993 for calcines, before the material was sent to the bin, are reported more extensively (Garcia, 1997: Tables 5-10), but some of the data columns are inconsistent with radioactive decay times. Analyses were not made for many components of interest, even in the feed solutions.

---

[3] These elements are present in various chemical forms. For the remainder of this report, the use of an elemental symbol does not necessarily imply that the element is present in the zero valent form. Typically, the elements are bound in compound forms, but the elemental composition, rather than the compounds, is of more fundamental importance to the processing steps under consideration.

TABLE 2-1 Physical Properties of Calcined High-Level Waste

Property	Value
Method of calcine formation	Fluidized-bed calciner
Temperature of formation	500 °C
Calciner startup material	$CaMg(CO_3)_2$ dolomite
Particle distribution	
Weight percent	Approximately 75% bed, 25% fines
Volume percent	Approximately 60% bed, 40% fines
Particle size	
Bed particles	200-500 μm diameter
Fines particles	200-10 μm diameter
Density	
Bulk	Approximately 1.4 g/cm^3
Bed particles	Approximately 2.7 g/cm^3
Fines particles	Approximately 0.9 g/cm^3
Attrition index	10-80% undisturbed
Thermal conductivity	
Alumina calcine	0.14 W/m-K at 38 °C
	0.52 W/m-K at 138 °C
Zirconia calcine	0.34 W/m-K at 600 °C
Heat Generation Rate	50-400 W/m^3
Fresh calcine	140-400 W/m^3
Angle of repose	30°-34° from vertical
Caking temperature	
Alumina calcine	$\geq$700 °C
Alumina-sodium blend (2:1)	>200 °C
Zirconia calcine	$\geq$700 °C
Zirconia-sodium blend (4:1)	Approximately 600 °C
Zirconia-sodium blend (2:1)	Approximately 400-600 °C
Zirconia-sodium blend (5:1)	Approximately 700 °C
Temperature Stability	
Zirconia calcine	
Volatilize residual NO_3	450-650 °C
Reduce NO_3 to 900 ppm in 15 min	660 °C
Volatilize Cs	$\geq$ 650-700 °C
Cs recondenses at	<600 °C
Volatilize Ru	>800 °C
Volatilize F	>900 °C
Alumina calcine	
Volatilize Ru	Approximately 800 °C
Volatilize Cs	Approximately 800 °C
Cs recondenses at	<600 °C
Volatilize Sr	1200 °C
Volatilize Ce	1200 °C
Volatilize B	1200 °C
Volatilize NO_3	$\leq$ 600 °C
Volatilize Hg	$\leq$ 600 °C
Sintering effects	
Zirconia calcine	Density increases by 45% at 1200 °C in 24 hr
Fluoride volatility at 900 °C	0%
at 1050 °C	4%
at 1200 °C	22%
at 1380 °C	85%
Not volatilized:	Ce, Pu, U

Table 2-1, Continued.

Property	Value
Sealed container pressure	
In 40 minutes for	
Zirconia-sodium blend	At 200 °C, 690 KPa max
(>5% residual NO_3)	At 500 °C, 1,630 KPa max
	At 700 °C, 1,700 KPa max
Zirconia calcine	At 600 °C, 460 KPa max
Alumina calcine	At 600 °C, 1,400 KPa max
Retrievability	
Alumina and Zirconia calcine	Retrievable by pneumatic suction after 8-10 years
Rate of retrieval	Avg. = 0.34 m^3/h

SOURCES: Table 1 of Garcia (1997), Berreth (1988), and Kimmel (1999c).

TABLE 2-2 Calcine Compositions Based on Analytical Results[a]

Element	Alumina Type Analytical Results (wt%)	Zirconia Type Analytical Results (wt%)
Na	0.97	4.51
K	0.21	0.46
Ca	14.5	22.8
F	3.16	12.2
Al	32.6	19.3
B	0.37	1.06
Fe	0.61	0.56
Zr	0.81	6.04
Oxygen and other components	46.77	33.07

[a]This table reports data on calcine types generated in 1993. These calcines were dissolved and the solutions analyzed to determine concentrations of calcine constituents.
SOURCE: Table 2 of Garcia (1997).

TABLE 2-3 Summary—Calcine Chemical Composition, Weight Percent Calculated from Flowsheets

	Al Calcine CSSF #1	ZR Calcine CSSF #4	CSSF #2	CSSF #3	CSSF #5	CSSF #6[b]
Al_2O_3	88.21	13.70	35.71	16.91	15.80	61.50
$Al_2(SO_4)_3$	2.34	0.00	0.00	0.00	0.00	0.00
$AlPO_4$	0.11	0.00	0.00	0.00	0.00	0.00
B_2O_3	0.68	2.35	1.45	1.99	2.26	0.44
$CaCO_3$	0.00	1.32	1.95	3.49	2.14	0.96
CaF_2	0.00	52.77	42.61	50.99	41.16	4.17
CaO	0.00	4.54	0.88	2.29	6.73	0.61
$Ca_3(PO_4)_2$	0.00	0.00	0.40	1.61	0.01	0.00
CdO	0.00	0.00	0.00	0.00	2.50	0.40
Cr_2O_3	0.00	0.38	0.26	0.45	0.12	0.12
Fe_2O_3	0.20	0.53	0.13	0.25	0.57	0.60
Gd_2O_3	0.00	0.01	0.00	0.00	0.00	0.00
HgO	3.04	0.05	0.91	0.06	0.19	0.17
$KAlO_2$	0.00	0.80	0.00	0.00	0.00	0.45
K_2SO_4	0.15	0.00	0.33	0.41	1.78	2.35
$MgCO_3$	0.00	1.11	1.64	2.94	1.90	0.81
MnO	0.00	0.00	0.00	0.01	0.02	0.17
MoO_3	0.00	0.00	0.00	0.00	0.00	0.02
$NaAlO_2$	0.00	1.09	0.36	0.57	0.87	22.02
$NaCl$	0.00	0.17	0.00	0.06	0.28	0.36
NaF	0.00	0.00	0.00	0.00	0.00	0.64
$NaNO_3$	3.94	5.61	2.06	2.83	10.56	2.09
$Na_3(PO_4)_2$	1.33	0.00	0.00	0.00	0.00	0.34
Na_2SO_4	0.00	0.00	0.16	0.12	1.96	0.28
Nb_2O_5	0.00	0.00	0.00	0.00	0.26	0.00
NiO	0.00	0.00	0.00	0.00	0.01	0.05
PbO	0.00	0.00	0.00	0.00	0.00	0.05
SnO_2	0.00	0.22	0.16	0.21	0.15	0.01
ZrO_2	0.00	15.33	10.99	14.79	10.73	1.39
Total	100.00	100.00	100.00	100.00	100.00	100.00
Total Volume, ft³	17,249	30,238	38,541	35,020	42,079	
Total Volume, m³	488	856	1,091	992	1,192	
Specific Gravity Density, g/cm³	1.20	1.60	1.44	1.59	1.54	1.20
Mass, kg	259,754	779,285	1,226,211	1,726,209	1,526,778	1,426,633

NOTE: Compositions for each CSSF for existing calcine were calculated as described in Garcia (1997: p.19).
SOURCE: Reproduced from Garcia (1990: Table 3f).

In general, the available data, analyzed for chemical, physical, and radiological properties, are not adequate for any type of calcine, for either process control or environmental considerations. Further, there appears to be no realistic sampling or characterization plan at this time to improve the database. However, it is necessary to know the approximate composition of the feedstock to be able to design applicable chemical processes. Representative sampling and analysis of the current bins is needed to determine the concentrations of the critical radioisotopes (i.e., Cs, Sr, and transuranic (TRU)) and other species of interest (i.e., those that would potentially interfere with radioisotope separations processes) in all of the bins.

A carefully defined sampling and limited characterization effort would significantly reduce the programmatic risk of inadequate retrieval. Data also are needed for (a) process testing, evaluation, and definition; and (b) environmental and regulatory purposes. To use limited resources most effectively, it is important to define what data are really needed and for what purpose, and then to tailor the requirements for each sample to the actual need.

RETRIEVAL, HANDLING, AND BLENDING OF CALCINE FROM BINS

Several successive calcining campaigns produced calcines of different compositions (e.g., alumina- and zirconia-based calcines, each differing in the relative abundance of aluminum, zirconium, and other elemental constituents such as calcium) that were injected into the bin sets. There appears to be a presumption that the calcines were deposited generally in layers with horizontal uniformity, as indicated by the common use of horizontal lines to delineate different calcine types in figures showing bin contents (as in Garcia, 1997: Figures 4b and 4c; see also Berreth, 1988: p. 2-1). However, different forms of calcine are probably commingled, because the pneumatic transfer systems deposited calcine in a cone at a significant and probably variable angle of repose (Garcia, 1997: p. 5), with inevitable avalanching of the conical deposits.[4] Therefore, it is unlikely that calcine of one composition (reflective of the liquid feed stock composition) could be retrieved separately from the other calcines deposited nearby, and the extent to which different calcines can be separately retrieved and segregated is questionable. Therefore, it would be unwise to develop processing specifications that assume the intact recovery and segregation of these distinct types of calcines.

Before processing or disposition of the calcine can occur, it must be retrieved from the bins and transferred to a holding tank that will provide feed to the process. The primary issue addressed here is whether or not any problems are likely to occur during such retrieval operations. A secondary matter is the possibility of using retrieval to blend calcines of different compositions to provide feed that is more amenable to processing than, for example, some variable sequences of compositions. Both issues are discussed below.

Retrieval Operations

Calcine is normally transferred from the calciner to the bins by entrainment (fluidization) in flowing air. Presumably, it can be transferred out of the bins in the same manner. This is a common and large-scale method used with dry particulate solids (e.g., in grain storage and transfer). It is expected to be successful so long as caking or sticking of the calcine does not occur. The question, then, is whether there are any mechanisms that can lead to cak-

---

[4] A "wall friction angle" of approximately 30 degrees is cited for actual aged calcine (Staples et al., 1979: Table XV, p. 18). Table 2.1 (Garcia, 1997: Table 1) lists 30-34 degrees as an "angle of repose."

ing. Two examples are (1) moisture entering the calcine, being absorbed by the hygroscopic components and causing the particles to stick together, and (2) a slow sintering bonding process.

The first caking mechanism, moisture effects, appears to be a possibility that would increase with time. The atmosphere in the bins should be protected from moisture, as by purging with dry air or protecting any openings to the outside with dryers.

The second caking mechanism, sintered bonding, is strongly influenced by calcine temperature. Although calcine temperatures are below normal sintering temperatures, sintering is a thermally activated process, which means that it could in principle occur over the long times (decades or more) that the calcine might be stored.[5] The effect of sintering, if any, would of course likely be concentrated in the thermally hottest part of the bins of calcine.

Should caking occur, it appears that this would not pose an unmanageable problem. It should be practical to break up the calcine by various mechanical means and then transfer it pneumatically. This would be a complication, but not likely a showstopper. An exception would be extreme caking, resulting, for example, from large amounts of water entering a bin. The committee believes this event to be unlikely. However, samples from many bins should be acquired to establish the capability of the calcine particles to flow freely.

The only data for production calcine retrieved from storage bins was obtained from two core samples, one of alumina- and one of zirconia-type calcines (Staples et al., 1979). The zirconia-type calcine had been stored for approximately 10 years with the maximum temperature decreasing from 220 °C initially to 190 °C. This material (13.2 kg of samples) flowed readily (i.e., it could be poured out of the sample tubes) and had a few small agglomerates. It contained <9 percent by weight fines (<200 mesh) near the bottom and about 3 percent by weight elsewhere. The sampler appeared clean when withdrawn. The alumina-type calcine had been stored for about 12 years with the temperature decreasing from 525 °C initially to 440 °C when sampled. This material (10.8 kg of samples) contained 18 wt percent fines and had "considerable cohesive strength" (Staples et al., 1979: p. iii). The sample tubes were coated with a visible film of dust, and resulted in air contamination. The calcine did not pour readily from the samplers, but required vigorous agitation and prodding. However, it was stated that this would not preclude pneumatic retrieval. Considerable dust was generated during sieving and pouring. There had been problems from plugging of transfer lines and the cyclone separator when this calcine was made. It is clear that this particular alumina-type calcine presented more potential problems than the zirconia-type calcine sample. The data show that calcine properties vary significantly, such that operational problems should be anticipated.

As reported in Staples et al. (1979), the force required for insertion of the sampler (a string of 28 sample tubes on a drill rig) did not cause a problem. There was some evidence for an easily penetrated crust at the top of the zirconia-type calcine but not the alumina-type. This may be significant since the zirconia-type contains hygroscopic substances. As stated above, the degree to which the calcine in bins is protected from atmospheric moisture was not made clear.

The quantity of calcine left in a bin after retrieval (the "heel") will depend on the nature of the calcine, the retrieval method, and the extent to which retrieval must be pursued. Certainly, as a minimum, there will be some fines left on the bin surfaces, and there might be

---

[5] Indeed, sintered bonding may be speculated based on analogy with the gamma alumina used as catalyst supports in the oil refining industry. These materials are made under conditions similar to Idaho National Engineering and Environmental Laboratory (INEEL) alumina calcines. The powder from fixed bed or fluidized bed heating is formed and sintered into monolithic structures at temperatures slightly above centerline temperatures in the bin. The gamma alumina, which enables high dissolution, an important attribute for recovery of the calcine, contains hydroxyl groups. Under the pressure of calcine weight to force interparticle contact, hydrogen bonds from mobile protons could provide the caking strength comparable to the strengths obtained for sintered catalyst supports.

considerable calcine that is difficult to retrieve for any of several reasons. The committee is not aware of any specification on the amount of calcine that must be retrieved from each bin, although the committee thinks that there should be one. For example, if a bin or tank is to be refilled with a Class A solid waste, or must meet some closure requirement, the residual calcine (left unretrieved) must not be in sufficient quantity to cause the appropriate limit to be exceeded. (This subject is treated in greater detail in Chapter 8.)

With only this limited database with real calcine, and in view of the variability of behavior between the two calcines examined, it would be difficult to conclude that there would be no problem with pneumatic retrieval. Indeed, the committee believes that there will be problems but that they can probably be handled. However, this eventually might require mechanical operations to aid particle flow and more elaborate retrieval methods (e.g., a manipulator arm) than simple pneumatic transfer. Clearly, characterization of additional calcine from other bins is required to better define this problem.

Calcine Handling and Blending

The concept presented to the committee for retrieval was vague with respect to how the calcine will be retrieved and blended to provide a somewhat uniform feed to the subsequent steps (whether treatment or direct solidification). Some degree of blending of different compositions of calcine will inevitably occur during retrieval, and it may be difficult to predict the extent of this blending or what part of the bin is being retrieved at any given time. Because calcine compositions are different, and the requirements for further treatment are different for different calcines, a considerable degree of feed uniformity will be required for efficient operation of most calcine treatment options. This can be accomplished by retrieving uniform calcine for feed (which may be difficult or impossible to assure), by retrieving a large volume into one or several bins in which it is mixed to provide a large quantity of feed of a reasonably uniform composition, or by similarly blending a large volume after dissolution. Blending to provide large volumes of uniform feed is commonly used in successful radioactive waste treatment operations.

Because some degree of blending is going to occur, it will be advantageous to design the system to benefit from this phenomenon, insofar as possible. Substantial storage capacity probably will be required (surge capacity or lag storage) to provide continuous availability of consistent feed. Indications from other DOE sites (and ongoing foreign radiochemical operations) are that such storage capacity is essential for operational reasons and for minimization of the final waste volume.

This blending raises the issue of whether it is feasible to sufficiently characterize the waste after retrieval from storage containers, rather than before. Subject to regulatory approvals, a "quasi-batch mode" such as the following may have to be adopted:

(1) Perform adequate testing on a range of surrogate compositions in order to understand the important process criteria and to develop suitable treatment requirements.

(2) Remove the waste in batches, homogenize to the degree necessary, characterize the batch, define the processing conditions, and then process the batch.

This "quasi-batch mode" might be slower (e.g., due to sampling and analysis requirements) and slightly more involved than a complete "front-end" characterization of the initial waste feed into a continuous process. However, for heterogeneous and potentially complicated processing, this quasi-batch mode might reduce the complexity of the treatment method or might improve the efficiency and reliability of operations. If this or another approach is adopted, some broad knowledge of the range of initial feed characteristics (i.e., in the current tanks and/or bins) would still be needed, because in order to fine-tune the chemical separations and/or immobilization processes, they must be developed to be adequate over the right range of conditions.

CALCINE DISSOLUTION

The basic dissolution approach proposed by INEEL is batch dissolution in nitric acid followed by use of settling tanks and filtration systems to remove undissolved solids (UDS). Data obtained from tests on a very limited number of samples suggest that as much as 98 percent dissolution of actual aged calcine might be achieved (assuming multiple (up to ten) recycles of UDS) under the following conditions: 5 M HNO_3, 10 ml acid/g calcine; 90 °C; and 30 to 60 minutes of contact time with heel recycle in each stage (shorter contact times also are suggested) (Brewer et al., 1997a: pp. 1, 7, 13; Fluor Daniel, Inc., 1997: p. 4-19). On blended calcine without recycle, dissolution greater than 95 percent was reported in some tests (Brewer et al., 1995). Dissolution for simulated calcine ranged between 71 and 97 percent (Herbst et al., 1995) and 46 and 95 percent (a function of stirring) (Brewer et al., 1997a). Accompanying the use of these dissolution values for planning purposes are the following assumptions:

- no calcine size reduction will be required prior to dissolution,
- no independent off-gas treatment system will be required,
- solid-liquid separation will be carried out by a combination of settling tanks and a filtration system yet to be defined,
- dissolver tank heels and filtered solids will be recycled to the batch dissolver tank, and
- residual UDS ultimately will be returned to the HLW stream.

Test parameters have varied in different experiments described in various technical reports. Table 2-4 below summarizes the dissolution performance that has been reported on three types of test materials: actual aged calcine retrieved from bins, radioactive calcine samples collected from the calciner during operations (without having been deposited in the bins), and simulants.

Although a high degree of dissolution (>98 percent) was observed in some of the experiments summarized in Table 2-4, these results were obtained for a limited number of tests. Indeed, the observation of much lower and highly variable dissolution percentages in other tests indicates that reliably achieving a high dissolution percentage in a large-scale process is questionable for the range of calcine feeds to be dealt with. The data cited in Brewer et al. (1997a), Herbst et al. (1995), and Brewer et al. (1995) demonstrate that a wide variety of dissolution challenges are to be expected. Therefore, large-scale dissolution tests on representative samples of typical (mixed) actual aged calcine, covering the full range of anticipated compositions, are needed to adequately demonstrate feasible dissolution conditions.

As Table 2-4 shows, the amount of UDS after acid leaching (and prior to any subsequent processing or separations) is uncertain. The only data with actual aged calcine gave 89 to 95 percent dissolution (Staples et al., 1979). Other tests involved selected calcine samples and therefore may not be indicative of the range of dissolution behaviors that will be encountered in practice. The committee concluded from Table 2-4 that a total UDS content of 1 to 10 percent is a reasonable estimate during full-scale calcine dissolution operations, with dissolution of Al calcine near 90 percent and dissolution of Zr calcine near 98, assuming ample contact time. Dissolution values of 90 to 99 percent result in a total volume of UDS of 63.5 to 635 m³ for processing approximately 3,000 m³ of zirconia-type and approximately 800 m³ of alumina-type calcine (Brewer et al., 1995).

Table 2-4 data infer that the 2 percent value for a UDS assumed by Fluor Daniel (1997) may be low. The data for actual aged calcine are extremely limited, and vary from around 11 percent UDS down to about 5 percent. UDS in the 1 to 2 percent range were reported for a few cases in which multiple batches of actual (but not aged) calcine were leached through the same dissolver (heel recycle). Multiple leaching of dissolution heels left in the dissolver appears to be somewhat effective, but the test data are too limited to draw a firm conclusion. It also appears that 98 percent total dissolution might be attained for most calcines using 5 M HNO_3 and heel recycle, although outliners clearly are to be expected as demonstrated by one test in which only about 50 percent alumina calcine dissolution was achieved. The unknown factor may be the amount of alpha alumina within various calcines, as this does not dissolve even after prolonged treatment. Unfortunately, the testing program has been too restricted to obtain data for a representative variety of the full range of calcine samples present in the bins.

Characteristics of UDS

While the dissolution data available to the committee are subject to considerable uncertainty for reasons outlined above, the data do suggest that parametric studies have identified the master variables of acid/calcine ratio, the final nitric acid concentration, contact time, and temperature.

The characterization data also strongly suggest that TRU constituents and strontium-90 (^{90}Sr) are preferentially retained in the UDS. For example, Table VI of Brewer et al. (1995) shows that Al-type calcine dissolved in the 78 to 96 percent range, while Am, Cm, and Pu are significantly enriched in the heels compared to the initial calcine. The UDS contains Al_2O_3, calcium-stabilized zirconia, and CaF_2. These results are consistent with the well known tendency of tri- and tetravalent actinides and Sr(II) to follow insoluble calcium and zirconia phases. However, not enough information is readily available to confidently predict the partitioning of actinides and ^{90}Sr between the UDS and solutions delivered to subsequent radiochemical separation. This gap in knowledge translates directly into uncertainty as to the ultimate fate of these key radionuclides in the flowsheet, including required capacities of the separations units and amount of HLW due to solid residuals.

While Brewer et al. (1995) includes a citation (reference 6) that reports that actinide buildup in UDS does not occur with heel recycle that report apparently was never published and this assertion therefore could not be evaluated.

Derivation of Dissolution Specifications

The degree to which the calcine can be dissolved affects the quantity and composition of the final waste form(s). The UDS component would probably be included in the high-activity fraction for immobilization because of its content of Sr and actinides. This high-activity fraction likely will have significant CaF_2 content coming either from solution or from

TABLE 2-4 Summary of Dissolution Data on INEEL Calcines

Report	Calcine Type	Test Parameters	Dissolution Results (expressed as a wt % dissolved)
Herbst et al., 1995[a]		2,5,8 M nitric acid, 90-100°C, 30 min contact time, and 10 ml/g acid/calcine ratio	
	4 Zr simulants		71-97
	Al simulant[b]		51-57
Brewer et al., 1995		all tests using 5 M acid, > 90°C, 10 ml/g, and contact times shown below	
	Al hot sample[c]	1.17 hours	78
		2.18 hours	85
		24 hours	91, 96
	Zr hot sample[c]	1.05–2.75 hours	95-98
Brewer et al., 1997 1998		5 M acid, 10 ml/g, >90°C, vigorous mixing, and sequential dissolution, each for 30-60 min	
	Zr hot sample[c]	10 sequential batches	98.7, 97.9
	Al hot sample[c]	5 sequential batches	98.8
	Zr simulant	5 levels of agitation	46-95
Staples et al., 1979		8 M acid, 10 ml/g, 90°C, 30 min	
	Actual aged Al[d]		95, 89, 95
	Actual aged Zr[e]		95, 93, 95.5
Slansky et al., 1977: pp 59-61		8 M acid, 5 ml/g, 95°C,	
	unspecified simulant	1 pass (30 min)	86.1
		2 passes (60 min total)	93.4
Slansky et al, 1978: pp 39-41		10 ml/g, 90°C, 30 min per pass	
	"Zr simulant"	1 pass, 8 M acid	85.8-86.9
		second pass at 8 M	92.4
		second stage at 12 M	89.5
	"radioactive sample"	8-10 M acid, 5-10 ml/g	94.5-96.6

[a] Table 1 of this report summarizes previous work that is not shown here.
[b] Gamma alumina was the dominant insoluble material in the test; the assumption given in Herbst et al. (1995), without readily available information to verify it, is that this form of alumina is absent in actual aged calcine.
[c] As indicated in these reports, the radioactive samples came from the calciner campaign "H3" in 1993.
[d] This is reported in Table XI of this reference.
[e] This is reported in Table XII of this reference.

UDS. If such waste streams with significant calcium or fluoride content is to be vitrified, a potential problem arises because borosilicate glass has limitations in incorporating constituents such as calcium fluoride (as discussed in Chapter 5). In this case, two remedies are readily identified:

1. modify the dissolution process to dissolve a larger fraction of CaF_2, which then would be removed from the HLW fraction in separation steps, or
2. produce an alternative glass (e.g., a phosphate glass) that is more tolerant of CaF_2 (this is treated more fully in Chapter 5).

The first remedy requires that dissolution requirements be derived based on considerations of the impact of undissolved components on the downstream processes.

ADEQUACY OF EXISTING INFORMATION

As stated earlier, the amount of characterization and testing data on actual calcine is inadequate for the selection of large-scale retrieval and treatment processes. Little direct information is available for actual aged calcine retrieved from bin sets. Conclusions on dissolution behavior are based largely on extrapolations from surrogate materials rather than the calcines currently existing in the bin sets. These surrogates and the few samples of actual aged calcine available for testing are not fully representative of all the calcine contents in all six bins. Additional characterization studies are needed to provide definitive conclusions about the full range of calcine characteristics and bounds on important calcine properties.

A second issue is that of the reliability of the current quantitative information on calcine characteristics, some of which comes from calculations that contain errors. Large fields of numbers in tables do not necessarily indicate an abundance of actual measured data. For example, Table 2-3 presents calculations based on data and flowsheet information (Garcia, 1997, p. 19). The "data" of tables 9 and 10 of Garcia (1997) are also calculated quantities, some of which contain errors. These tables give isotopic abundances after 500 years for many isotopes present in the calcine. For some of the isotopes of critical importance, the committee recalculated these abundances using straightforward half-life decay methods to confirm their accuracy. Many were found to be in error. Some examples are the values for ^{238}Pu, ^{241}Pu, ^{244}Cm, ^{137}Cs, ^{151}Sm, and ^{166m}Ho in Table 9, and ^{210}Pb and ^{226}Ra in Table 10 of Garcia (1997).

In other key reports, the quality of analytical data that is available is clearly inconsistent and contradictory. For example, Tables 4.3-3 and I-17.2 in Fluor Daniel-Hanford, Inc. (1997), which describe average alumina and zirconia content in the six bin sets, are in conflict. Bin set numbering appears to be scrambled for bin sets 2 through 4, and the zirconia content is listed as 0 percent ZrO_2 in one case and 10 percent Zr in the other. Another example of significant analytical inconsistency is found in Table 11 of Brewer et al. (1997a), where radiochemical analyses of UDS from a zirconia calcine dissolution test are listed. In this table, the gross alpha content is listed as 7.1×10^8 dps/g, yet the sum of the individual TRU constituents is approximately four orders of magnitude less. A similar discrepancy exists in Tables 13 and 15 of Brewer et al. (1997a). While there may be reasonable explanations for such discrepancies, they are not explained in the reports. The committee therefore is unwilling to base firm conclusions on such questionable analytical data.[6]

---

[6] To verify the accuracy of analytical chemistry results, one possible approach used by some DOE laboratories has been to provide confirmatory samples to another DOE laboratory (with suitable expertise) for independent testing. Examples of laboratories with sufficient expertise include the Isotope Divisions of the Los Alamos

CRITICAL TESTING NEEDS

The committee believes that substantial and sustained testing with a full range of actual aged calcine spanning all production variations is required to provide adequate confidence in the understanding of partitioning of radioactive constituents between UDS and liquids. Even more critical is the need to demonstrate that solid-liquid separations (Chapter 3) will be adequate to meet the demanding separation factors required by the proposed flowsheets, particularly in meeting Class A waste levels. The key question, which is one of judgment as well as adequacy of resources devoted to the problem, is whether the required information can be obtained in parallel with design efforts, as suggested in Fluor Daniel, Inc. (1997). Because of the expense and difficulty in obtaining the required bin set materials, the judgment of the committee is that, even if adequate resources are provided, it is unlikely that the required information can be obtained in the timeframe specified by the baseline plan.

This recommendation for further testing regarding dissolution is consistent with conclusions of earlier external review groups, as listed in Appendix G of Murphy et al. (1995). In addition, the reports the committee reviewed repeatedly point out the requirement for additional testing. For example, Table 12.3-1 in Fluor Daniel, Inc. (1997) recommends as a high priority a 2-year dissolution study with actual aged calcine material, and recommends that this be carried out prior to early conceptual design. As another example, Brewer et al. (1997a) states that additional testing of the quantity and nature of radionuclides in UDS must be carried out with actual calcine. The committee concurs that the tests should include dissolution rates, temperature, acid requirements, end-point acid concentration, and residue characterization as well as settling/filtration tests.

Chapter 13 presents recommendations based on the material in this chapter. As discussed above, (1) the characterization data on actual aged calcine are limited, and (2) testing of planned operations should be done on materials that span the range of the different properties of calcine that are stored in the bin sets. These characterization, retrieval, blending, and dissolution operations are important, insofar as they will determine the ultimate fate of key radionuclide constituents. That is, those radionuclides that are not retrieved are left in situ, and those that are dissolved and partitioned from the high-activity fraction are the only candidates for a possible low-activity fraction.

National Laboratory and the Lawrence Livermore National Laboratory, which conduct actinide and fission product diagnostics chemistry on nuclear debris. If the analytical chemistry data are not a problem, but instead the issue is the use of such data in calculations and the presentation of results in technical reports, this latter issue can be resolved with appropriate quality assurance procedures.

3

Physical and Chemical Separations

On a weight basis, most of the calcine is nonradioactive. Therefore, separation of the less radioactive constituents into a low-level waste stream has the potential to reduce the volume of high-level waste (HLW) to be disposed. Various separations objectives[1] [e.g., separation of transuranic (TRU) components only, or separation of TRU, cesium (Cs), and strontium (Sr) components from the remainder of the waste] have been proposed in U.S. Department of Energy (DOE) literature. All of the options require calcine dissolution followed by separation of solids from liquids and additional chemical separation steps on the liquid fraction.

This chapter discusses general separations strategies and specific separations methods that could be used to develop waste fractions of high and low activity. Physical (solid–liquid) separations are treated first, followed by chemical separations and detailed discussion of cesium, strontium, and TRU separation plans.

THE SEPARATIONS APPROACH

The calcines stored at Idaho National Engineering and Environmental Laboratory (INEEL) contain large quantities of nonradioactive constituents from the fuel assemblies[2] and from materials added during fuel reprocessing and waste conditioning and calcination. Because a high weight and volume fraction of the calcine is nonradioactive, separation of radionuclides from this bulk is a way to reduce the volume of HLW to be immobilized prior to disposal. If the costs of separations process steps were not excessive and the risks acceptable, this strategy might reduce overall disposal costs and hazards. A full separations approach could remove all of the long-lived actinides, plus the ^{137}Cs and the ^{90}Sr. These are the critical elements whose removal might then allow the remaining material of reduced activity to be delisted from further management as HLW.[3] The residues remaining after this partitioning could fall into the

---

[1] Another important separations objective is the segregation of hazardous chemical constituents [e.g., mercury (Hg) and lead (Pb)] from nonhazardous components. This issue is discussed in a limited way in the context of each separation option of this chapter, and more broadly in Chapters 9 and 10.
[2] Some of these initially nonradioactive fuel assembly materials become radioactive during reactor operations. For example, in the initially nonradioactive zirconium cladding material, Zr-93, a long-lived beta (β) emitter, is generated during exposure to radiation inside nuclear reactors.
[3] This reclassification from HLW to some other category may require regulatory approval, most probably from the USNRC.

U.S. Nuclear Regulatory Commission (USNRC) waste Class A category, possibly enabling shallow land disposal of the materials in a suitable land disposal facility.

Unfortunately, as described in reports cited in this chapter, chemicals present in the current calcines interfere with the extraction processes proposed for partitioning. This complicates the chemistry of proposed processes and may require additional "head-end" treatment to remove the interfering elements. The extra chemical treatment could expand facilities needed to perform this work and significantly increase overall disposal costs. The merit of a separations approach is affected by these considerations, which are treated in more detail below.

Comments are offered on cesium ion exchange to separate cesium, the strontium extraction (SREX) process to separate strontium, and the transuranium extraction (TRUEX) process to separate actinides, or TRU elements. Application of one or more of these separations processes is under consideration for the INEEL HLW.

SOLID–LIQUID SEPARATIONS

An efficient solid–liquid separation (SLS) process is necessary to assure process operability and to allow high decontamination factors (DFs)[4] to be achieved. Many events can complicate downstream operation without completely preventing it; however, a few complications, like flow blockage caused by accumulation of solids resulting from inadequate SLS, or leaks due to corrosion or equipment failures, can result in unplanned plant shutdowns. The SLS process is therefore one of the critical, if not the most critical, operations. Radiochemical processing experience provides adequate examples (e.g., plugged transfer lines) in which solids in process streams caused a facility to be inoperable or a process to perform unacceptably. The committee believes the importance of SLS is not sufficiently recognized; therefore SLS has received inadequate consideration in both design and development work. To reduce the technical risk, it is recommended that serious development studies be conducted with at least two different SLS approaches using dissolver solution produced from actual aged calcine (i.e., solution derived by leaching calcine with nitric acid, as discussed in Chapter 2, to dissolve calcine components into nitrate forms). Work with simulants is not appropriate for this purpose.

The most significant SLS problem is likely to occur in clarifying the feed solution following calcine leaching. Unusually good feed clarification is especially important when large DFs are required, as in the case of USNRC Class A separations. Because the feed solution is almost certainly saturated with respect to critical chemical constituents of the heel, and chemical conditions are changed during the several subsequent operations, there is a strong probability that new solids will be generated during processing (as noted in some tests). Thus, more than one SLS operation may be needed.

Two important reasons for requiring an effective SLS to deal with this feed clarification problem follow:

1. The residual solids are enriched in Sr and TRU, the limiting components in decontamination operations to produce Class A waste. Inadequate solids removal has an effect analogous to that of an incomplete separation process. Small-particle solids (in the sub-micron and perhaps colloidal range) are the most difficult to remove and commonly carry TRU preferentially. Therefore, efficient removal of solids down to small-particle size is required to achieve a high DF.

An additional crucial consideration that apparently has not been adequately addressed is the role of colloids, which are notorious in plutonium chemistry. Because colloids are un-

---

[4] The decontamination factor, defined in Appendix E, expresses the degree of separations.

likely to be removed by the proposed filtration systems and can cause significant degradation of downstream separations operations, they potentially can negate the benefit of otherwise outstanding aqueous separations systems. Potential problems derived from colloid behavior in calcine processing will be exceedingly difficult, if not impossible, to evaluate without testing with an adequate range of actual aged calcine.

One report (Brewer et al., 1995: Appendix D) addressed the radioactivity limits in zirconia calcine undissolved solids (UDS) with respect to USNRC Class A requirements. This report concluded that 97.8 to 99.9 percent solids removal is necessary for transuranics in solid (grout) waste prepared from clarified supernate, and 99.9 to 99.99 percent UDS removal is needed for ^{90}Sr. Moreover, because of the "sum of the fractions" rule (see Appendix E), the usual variations in process performance, and additional contributions to Class A limits from soluble constituents (e.g., Tc and residual Cs, Sr, and TRU), the specification for solids removal should be about a hundred times larger than that required for the most limiting contributor to meet the Class A limit. That is, clarified feed should have solids removed by a factor greater than 10^5. It appears that the significance of this requirement has not been recognized.

2. Physical operability of downstream treatment equipment, such as ion exchange columns or solvent extraction banks, generally requires highly clarified feed. Ion exchange columns retain most solids, and this will block flow if enough solids are present. Ion exchange columns are sometimes successfully used as, in effect, deep-bed filters if the combination of throughput and feed solids content is small enough. This can be advantageous in a once-through process, which may be required in any case. However, the design studies propose to operate ion exchange media through multiple elution and loading cycles; and experience has shown that contaminated solids associated with the ion exchange bed result in degraded decontamination in subsequent cycles. When multiple ion exchange load-elute cycles are used, the presence of even small amounts of solids in the feed generally causes lower raffinate DFs, lower column capacity, and increased pressure drop or slower flow, all of which degrade performance.

Although some solvent extraction contactor equipment will operate physically with solids present in the feed, other contactors that may accumulate solids are more sensitive and may become inoperable. The degree of chemical separation of radionuclides (e.g., TRU and Sr isotopes) from these solids is generally degraded seriously. Solids deposited in process equipment become a subsequent source of radioactivity leaching into downstream processes. A more serious problem results from accumulation of solids at the phase interface, which not only decreases DFs, but also can seriously interfere with phase separation and system operability. The DF requirements of some of the process steps will be difficult to achieve at best; therefore, anything that might degrade performance is not acceptable.

Current Status

As stated above, the information provided to the committee indicates that SLS has not been adequately addressed. There is little quantitative data other than on total UDS and none on UDS particle size and other characteristics critical to SLS. For the most part, settling followed by some sort of filtration seems to be assumed in the information provided to the committee. For example, a design study (Fluor Daniel, Inc., 1997) uses the following standard description in the two applications—SBW and calcine leach solution—where SLS is considered:

"Ultimate filter sizing and type selection should be based on actual performance testing. It is anticipated that the mean particle size is less than 0.5 micron, and polishing [final] filtration is required to reduce particulate loading prior to separations processing of the SBW [or dissolved calcine]). Evaluation of cartridge filters to effect this polishing filtration should be made with an additional selection criterion being ultimate disposal in vitrification."

This description says essentially nothing about the requirements or the method for SLS. The same report assigns "filtration" testing a medium priority, with less than a two person-year effort, which the committee believes is inadequate.

Recent reports on TRUEX processing propose to change the dissolution procedure by reducing the initial and final nitric acid concentration (Herbst et al., 1998; Brewer et al., 1997a). The reason relates to improvements in both the operability and performance of the TRUEX process (see below for further discussion). However, this change would probably affect the extent of dissolution and the nature of the UDS, and it is even more likely to increase the fraction of Sr and TRU that does not dissolve. These variables are important to SLS. A change such as this affects many unit operations (in fact, the entire flowsheet), not just the one that instigated the change. The interdependence of the unit operations on changes in conditions for any one operation must not be neglected.

In various publications, the filter type discussed appears to be either a cartridge or a back-washable filter, or both, but is not described clearly. In the approaches examined, there is high throughput with a significant fraction of solids that must be recovered. An estimate (a non-conservative one, in the committee's view) of 63.5 to 635 m^3 is given in Table XVIII of Brewer et al. (1995); at the same time, the acceptable UDS content in the filtrate is extremely low. Meeting both of these conditions will not be easy. In the committee's experience, treating waste requiring high DFs suggests that such straightforward clarification schemes are often inadequate. A result has been the use, in practice, either of large deep-bed filters (which is disadvantageous because the deep-bed material is routed to HLW and thus increases the solids to be immobilized as HLW) or of cross-flow membrane filtration with very small pore-sized media, such as ultrafiltration (which is disadvantageous because of increased operational complexity and cost).

Clarification is perhaps as much an art as a science, but if feed variability is anticipated, membrane filtration is most likely to be successful. During the course of this review, a report on cross-flow filtration testing was provided (Mann and Todd, 1998) that indicates an interest in this approach. The testing did achieve moderately low UDS; however, the study was limited to a single membrane filter (0.5 μm pore size), and results were not fully encouraging because of rapid flow reduction, probably the results of particles being trapped inside the pores. A smaller pore size will likely be required for cross-flow filtration because there is little or no filter cake to protect the filter media from being clogged by small particles. The committee recommends that increased priority be given to SLS, with evaluation of at least two different types of filtration, one of which should be cross-flow filtration. The use of simulants is of little if any value for such a study, since the actual solids in a variety of real aged wastes must be addressed.

CESIUM ION EXCHANGE SEPARATION

There are only a few options for recovering Cs from acid solutions. Most options involve either coprecipitation or inorganic ion exchange. A multistage continuous process using ion exchange columns is advantageous because it can result in the required extent of separation with considerably less sorbent than would be needed for one or a few batch separation stages, for a given sorbent (and distribution coefficient). Although an ion exchange material in pure powder form gives the best exchange properties, it usually cannot be used in a column because of significant flow resistance. In principle, the powder can be granulated with a binder to yield larger particles suitable for column use. The performance of granulated particles, however, is generally degraded because of the longer diffusion paths, blocking of pores, and changes in properties (with silica-based binders) caused by treatment at elevated temperatures to set the binder.

The most selective sorbent materials for acid-side Cs separations are not widely available in the durable granular form required for column use. Some granular sorbents are currently available in research and small production quantities, but there is no assurance that they will be produced commercially over the timeframe of interest. Sorbents can usually be prepared easily in a powder (slurry) form. If the distribution coefficient is large enough (i.e., in excess of 10^4), the sorbent can be used in a batch mode with one or possibly a few batch separation stages. As the distribution coefficient increases, less sorbent is required. Such batch operation (which is not as dependent on specific material availability but would generate more waste sorbent) has apparently not been considered seriously. Many sorbents can be easily prepared in a slurry or powder form and used in a batch mode.

Limited information was provided for the following three granular Cs sorbents considered in this program: FS-2 (a Russian-prepared potassium copper hexacyanoferrate granulated with a silica binder), AMP-PAN (ammonium molybdophosphate granulated with polyacrylonitrile and prepared by the Czech Technical University, Prague), and IONSIV IE-911 (a crystalline silicotitanate developed at the Sandia National Laboratory and commercialized in a granular form by Universal Oil Products). Only the last sorbent is clearly available in quantity. The first two have limited performance data. The literature base for nongranulated sorbents is much more extensive than for these granulated forms, although it is still rather small for acid systems.

Experimental Basis

The FS-2 evaluation was based on limited testing, first in Russia and later at INEEL. Batch distribution measurements and one column test in Russia, using SBW simulant, gave promising results that led to a decision in 1997 to switch the reference process from AMP-PAN to FS-2 (Olson, 1997). Additional (but still quite limited) tests at INEEL, using both dissolved calcine simulants and real SBW, gave highly varied results (Todd et al., in press; Brewer et al., 1996). Loading results differed with different feeds, possibly because of interference of some constituents (e.g., mercury). Column breakthrough occurred much earlier with real SBW feed than with simulants in column tests. These results led to a recent decision to switch the sorbent choice back to AMP-PAN.

In principle, the FS-2 sorbent can be eluted with strong nitric acid and regenerated for repeated loading cycles, thereby generating relatively little waste as spent sorbent. Although the available data suggests that substantial elution is possible, the required high degree of elution for Class A separations will be difficult to achieve. Thus the committee believes that, in practice, the required large DF in the treated column effluent probably cannot be reliably obtained after the first loading cycle, because residual Cs on the column after elution will bleed into the product during subsequent loading cycles due to a reversal of the ion exchange absorption reaction, a reversible equilibrium. Therefore, the committee considers the assumption of adequate elution of this sorbent to be non-conservative. The Fluor Daniel, Inc. (1997) design study assumed the use of FS-2 for repeated cycles with a life of 1 year. Such a long lifetime is undocumented, and the commercial availability of the material is uncertain. This information demonstrates that the existing database is inadequate to support selection of FS-2 for Cs removal.

Other hexacyanoferrates may be practical for Cs removal. In particular, potassium cobalt hexacyanoferrate has been used in Finland on a significant scale for treating neutral and alkaline reactor waste (Harjula et al., 1994). This material can be made in a granular form directly (no binder and granulation process), and was once marketed in the U.S. by BIO-RAD. It has been shown to be useful also in acid solutions.

AMP-PAN was evaluated in tests at INEEL using both simulants and dissolved calcine (Miller et al., 1997). The results were interpreted favorably because of large distribution coefficient values (ca. 3,000 ml/g) measured for actual SBW and dissolved Zr calcine. Unfortunately, similar experiments yielded distribution coefficients about sevenfold smaller with simulants of these two wastes, and even smaller values with both actual and simulated Al calcine. Because simulants should, by design, faithfully represent the requisite properties of actual aged waste, these data must be considered troublesome unless the discrepancy can be explained and reproducible data can be demonstrated. The smaller range of distribution coefficients would degrade performance with respect to column capacity, resulting in increased generation of HLW solids.

Two column tests were conducted with AMP-PAN, one with actual alumina type calcine and one with a calcine simulant (Todd et al., in press). Each column was eluted with ammonium nitrate solution and loaded for a second cycle. In both cases, the second loading cycle showed a higher Cs concentration in the initial effluent and an earlier breakthrough. It is unlikely that adequate decontamination can be achieved after elution of AMP (as is also the case for FS-2), in which case a once-through process would be required; this results in increased solid HLW because the loaded sorbent becomes HLW. An optimistic estimate based on loading 3,000 bed volumes in a once-through mode suggests that at least 27,000 kg of sorbent would be required. In Todd et al. (in press), literature on the use of AMP in acidic systems is not adequately referenced.

AMP-PAN particles are agglomerated with a polyacrylonitrile organic binder (the FS-2 and IONSIV IE-911 sorbents use inorganic binders). The radiolytic stability of this material, with respect to both aggregate degradation and generation of flammable and toxic gases, apparently has not been demonstrated. Thus, even if elution should prove to be practical, the lifetime of the sorbent is not known.

IONSIV IE-911 was also tested (Todd et al., in press). The Cs loading capacity was smaller than with FS-2 or AMP-PAN by a factor of two or more. It was found that, for all practical purposes, the IONSIV IE-911 sorbent could not be eluted. This would imply that the loaded sorbent would be part of the HLW fraction, thereby increasing the solid HLW volume. The committee suspects that inadequate elution is a feature common to all the inorganic sorbents as long as raffinate decontamination to well below the USNRC Class A limit is required. IE-911 has one distinct advantage in that it is the only one that is definitely commercially available.

Current Status

The three sorbents discussed above (FS-2, AMP-PAN, and IONSIV IE-911) have undergone limited comparative testing with radioactive calcine samples (TFA, 1998). With FS-2, the presence of mercury significantly reduced the sorbent's capacity for Cs, and therefore FS-2 was dropped from further consideration. AMP-PAN demonstrated excellent selectivity and capacity and currently is selected as the baseline technology for the full-treatment option. IONSIV IE-911 currently is selected for the limited treatment option in which Cs is stored indefinitely on the sorbent. AMP-PAN was not selected for the latter option because of the organic binder, which could be unstable during storage.

The committee concludes that because the Cs sorbent of choice has changed from AMP-PAN to FS-2 and back to AMP-PAN within the last two years, significant uncertainties exist regarding sorbent selection. Much more definitive research is required with a more representative range of feed materials and a wider range of experimental parameters. The current database[5] is not adequate to justify a selection. Certain key criteria (such as the required DF for the aqueous product, practicality of adequate elution to permit multiple cycles, sorbent life, quantity of spent sorbent destined for the HLW fraction, and compatibility of the sorbent with the immobilization process) are not known well enough to permit an adequate analysis of options, definition of a flowsheet, or evaluation of a process.

Several other Cs separation options were not considered in the detailed written information provided to the committee. One is a cobalt dicarbolide sorbent, which apparently is being applied on a large scale in Russia. Another is the use of a few (e.g., three) stages of simple batch contactors/settlers using inorganic sorbent powders (such as several hexacyanoferrates or AMP), which are easily prepared, followed by good SLS (as in Campbell and Lee, 1991; Campbell, et al., 1991). Batch contactors would generate more solid sorbent to add to the HLW, compared to column operation, but this approach avoids dependence on the foreign commercial availability of expensive granular sorbents for which the long-term supply is questionable.

STRONTIUM SEPARATION

Based on the information provided (e.g., Garcia, 1997), calcine presently contains, on average, approximately, 5 Ci of ^{90}Sr per kg of calcine. Assuming a density of 1.2 to 1.6 g/cm^3 for calcine (Garcia, 1997: Table 3f), this yields 6,000 to 8,000 Ci/m^3. The Class A and C limits of the USNRC's 10 CFR 61 for ^{90}Sr can be met with radioactivity concentrations of 0.04 Ci/m^3 and 7,000 Ci/m^3, respectively (Fluor Daniel, Inc., 1997: Table 3.2-3). Therefore, no ^{90}Sr separations are required to meet Class C limits if the "sum of the fractions" rule is ignored.[6] However, extraordinarily high separation factors for ^{90}Sr are required to achieve Class A levels. In the briefings received by the committee, INEEL staff cited ^{90}Sr separation factors of 14,000 to 45,000 to meet Class A limits. Much lower Class A separation factors of 1,300 to 4,000 are required for the less radioactive SBW. According to Raytheon Engineers & Constructors (1994: p. 2-12), a DF of 10^5 is required for converting dissolved calcine to Class A waste, using the assumption that this DF can be achieved with two strontium extraction (SREX) process cycles. Although these references do not contain the complete derivation of these DF values, the large values indicate high separations requirements that in practice can be very difficult to achieve, particularly when there is significant uncertainty in the variability of the feed composition.

A number of prior review groups (Murphy et al., 1995: Appendix G) have concluded that the SREX is the preferred method for removing ^{90}Sr from the acidic SBW or dissolved calcine solutions. SREX uses a derivative of 18-crown-6 ether dissolved in a hydrocarbon solvent as extractant to remove strontium from acidic solution. Tributyl phosphate (TBP) is added as a phase modifier. General characteristics of SREX are relatively low single-stage

---

[5] Thompson (no date given) provides a recent update of the current state of development of several cesium separation methods.
[6] In support of this statement, although the projected average ^{90}Sr activity is close to the Class C limit in this calculation, this activity would diminish over time due to radioactive decay, and blending will tend to reduce any areas of localized activity greater than the average.

distribution coefficients for Sr (with K_d values of 3 to 5) and significant extraction of Pb, Hg, TRU, Ca, Na, and K. Pb and Hg [both being metals whose disposal is covered by the Resource Conservation and Recovery Act (RCRA)] are especially troublesome for SREX because it is difficult to strip them from the organic extractant. Stripping difficulties can seriously threaten the ability to achieve the high separations required for Class A product. RCRA requirements regarding Pb and Hg can significantly impact waste disposal options. Na, K, and Ca are potential interferences because of mass action effects (Ca concentration can be 10^6 times the Sr concentration). As a result of these factors, and the relatively low SREX distribution coefficients for Sr, high Sr separations factors can only be achieved with multistage separations and careful control of operating conditions.

A feasibility study report (Fluor Daniel, Inc., 1997: Section 4.5.8) contains a concise description of the SREX process as applied to SBW and dissolved calcine. To minimize complications from undesired coextracted cations (notably mercury and TRU); SREX would follow Cs ion exchange and TRUEX. Countercurrent solvent extraction processes using up to 24 stages of centrifugal contactors have been proposed. Strontium (and notably Pb) is stripped with 0.01-M aqueous ammonium nitrate solution. This strip solution would be added to the HLW stream, and the solvent would be chemically purified and recycled to the contactors.

Chapter 4 and Table I-13.1 of the feasibility study (Fluor Daniel, Inc. 1997) describe anticipated operating parameters for SREX. This study lists "apparent separation factors"[7] as 8.44×10^{-10} (Sr) and 2.54×10^{-4} (Pb) for SBW and 1.42×10^{-9} (Sr). These factors are most likely derived from a theoretical calculation of separations achieved across many successive stages. In practice, the single-stage separation factors in multistage processes decrease drastically through the cascade to essentially one as the target species becomes more dilute. This effect usually prevents the achievement of the theoretical value that is based on a given number of stages. Therefore, a 10^{-10} "apparent separation factor" is, in the committee's judgment, an unrealistically optimistic estimate for an integrated process, even with two cycles as proposed. The basis for calculating the "apparent separation factor" is not given in the Fluor Daniel, Inc. (1997) study, which does not explicitly include all the input parameters and assumptions used (Kimmel, 1999b). These calculations depend on normalizations that arise from the fact that other solution inputs (e.g., scrub flows) involve feed and decontaminated raffinate solutions that differ in volume, providing dilution of Sr irrespective of any separations achieved in the extraction stages. Nevertheless, as with the DF values quoted at the beginning of this section, these "apparent separation factors" represent stringent separations requirements.

Adequacy of Existing Information

INEEL technical reports published in the past few years form much of the technical basis for the site's selection of SREX for ^{90}Sr removal. The major findings of these reports are briefly summarized below.

According to Raytheon Engineers & Constructors (1994: p. 4-23), an overall DF of 4,500 was obtained in three successive batch treatments of "actual nuclear waste solution." The report adds that the SREX process had been tested on a bench-scale with simulants, but had not been demonstrated with solutions using actual aged calcine waste.

Law et al. (1996) describes centrifugal contactor studies on simulated SBW containing radioactive Sr and Pb tracers. This study indicated that 99.98 percent Sr removal was achieved. Inadequate nitric acid stripping of Pb was observed, leading to concentration of Pb

---

[7] In the ensuing discussion, the term "separations factor," defined in Appendix E, is used to express the degree of separations. Differences in the usage of this term exist among practitioners; in the feasibility study of Fluor-Daniel, Inc. (1997), the term "apparent separation factors" is used.

in the organic solvent and precipitation of a Pb compound. Aqueous ammonium citrate was shown to effectively strip both Sr and Pb and eliminate Pb precipitation. Stripping efficiencies of 95 percent for Hg, 63 percent for Zr, 8 percent for Al, 9 percent for Ca, and 11 percent for Na were achieved.

Law et al. (1997) describes centrifugal contactor studies with simulated and actual SBW. Twenty-four stages of separation using 2-cm contactors led to >99.995 percent ^{90}Sr removal and >94 percent TRU removal, stated to be sufficient to achieve Class A levels. In addition, >94 percent Pb and 83 percent Hg were extracted, but only 5 percent of the Hg was stripped from the organic phase with 3 M nitric acid, meaning that essentially all of the Hg remained in the stripped SREX solvent stream. In practice, high Hg retention would not be acceptable since solvent reuse would not be possible in this process system.

In Wood et al. (1997), batch extraction experiments with simulated and actual SBW and dissolved actual calcine are described. Pb, Hg, Na, K, U, and Pu were effectively extracted as well as Sr. The actinide extraction was attributed to the TBP rather than the crown ether. Because Sr back extraction is only slightly acid dependent, interferences from alkali metals appear to be manageable. While Hg was found to be efficiently extracted, stripping from the organic phase was not successful, and the report recommended additional work on Hg stripping. Precipitates proposed as crown:Pb:nitrate complexes were observed at the organic/aqueous interface.

Critical Testing Needs

The committee believes that the demanding ^{90}Sr separations required to achieve Class A criteria, combined with difficulties indicated by the limited amount of testing carried out so far, represent high programmatic risk. The committee concurs emphatically with the technical reports that repeatedly make the case for additional development. For example, in Chapter 4 of Fluor Daniel, Inc. (1997) it is stated that "Prior to the actual design of a full-scale plant, additional engineering data and complete performance testing of a recycling solvent system will be required." As a second example, the Raytheon Engineers & Constructors (1994) study states that "Laboratory and pilot plant testing with simulated and actual dissolved calcine is necessary for complete development of a SREX flowsheet" (p. 4-23).

Elsewhere in Fluor Daniel, Inc. (1997), it is recommended that the same type of additional engineering and operations data is required for both TRUEX and SREX. Specifically, complete subsystems including extraction, scrub, strip, wash, and acid rinse contactors need to be constructed and subjected to sustained tests. Critical process vulnerabilities to be investigated include solvent/extractant recycle and degradation, impurity buildup in the organic phase, temperature effects, and formation of precipitates and emulsions.

The behavior of the RCRA-regulated constituents Pb and Hg are particular concerns warranting additional investigation with actual aged calcines because of the difficulty in stripping them from the organic phase, the possible formation of precipitates, and uncertainties in calcine properties. Hg also is of special concern because removal of Hg in the upstream TRUEX operation is intended (and this step, too, is problematic, as noted below in the TRUEX section).

Nowhere is the need for excellent upstream SLS more critical than in Class A SREX separations, where high decontamination factors are paramount. Whether expressed as "apparent separation factors" or as DF values, the numbers cited previously for Sr removal indicate stringent separations requirements to meet Class A levels.

Although the committee concurs with previous review groups that SREX is a promising approach for removing ^{90}Sr from SBW and dissolved calcine, the committee believes that the current technical status of SREX in application to INEEL calcine, and to a lesser extent

SBW, is far too immature to risk the multibillion dollar investment that would be required to build an integrated Class A separations plant. Additional SREX testing with actual aged calcine clearly is required before committing to this action.

Alternatives

Other alternatives such as inorganic ion exchange may be feasible. It is unlikely, however, that the high DFs required to meet Class A limits could be achieved without generating an unacceptable quantity of additional HLW in the form of spent sorbent. Other activities in pursuit of alternative separations techniques, such as solvent extraction based on chlorinated cobalt dicarbollide (TFA, 1999), were not presented in sufficient detail to the committee to form a position on these methods.

TRUEX SEPARATIONS PROCESS

The TRUEX chemical process for extracting TRU elements (i.e., actinides) from an aqueous media into an organic media uses a strong chelating agent dissolved in an aqueous-immiscible organic diluent. In TRUEX, the primary chelating agent is octyl (phenyl)-N, N-diisobutylcarbamoyl methyl phosphine oxide (CMPO) blended with TBP, both dissolved in an inert aliphatic diluent similar to kerosene. The TBP is included to prevent third-phase formation when the chelating agent is heavily loaded with extractable ions. Although the system is similar to PUREX (TBP alone in an immiscible organic solvent which is used to extract only U, Np, and Pu from fission products in nitric acid solutions), TRUEX will readily extract the +3 and +4 actinides as well as the uranyl, neptunyl, and plutonyl ions. A disadvantage is that it also extracts +3 lanthanides and several other common +3 and +4 elements (Fe, Zr, etc.) as well as the Hg present in INEEL wastes.

The utilization of TRUEX and PUREX processing systems has been enhanced by the advent of compact, high-throughput, centrifugal contactors. These devices are highly efficient when used as single-stage extractors and are specifically designed to link into extended cascades. This capability permits low distribution coefficient extractions to be linked into cascades of many stages to produce high separation factors for chemical purification. Unfortunately these devices, which use high-speed rotors to generate the interphase contact required, are not tolerant of precipitates formed during separations. Therefore, feedstocks and process reagents must be chosen to prevent formation of interstage precipitates during operation. This requirement becomes critical when dealing with very complex feedstocks such as the dissolved INEEL calcines that contain large amounts of fluorides, aluminum salts, and zirconium compounds.

The idea of simply stripping the actinides from dissolved calcine by feeding it to a TRUEX contactor cascade is an inviting concept, but the realities of making such a process work over the wide range of chemical compositions in these complex feeds will probably lead to major difficulties, as discussed below.

Technical Problems

As discussed in Calcine Dissolution in Chapter 2, the dissolver product must be clarified prior to chemical processing. This UDS residue will be enriched considerably in actinides and ^{90}Sr in a complex mixture of chemical compounds that would include Al_2O_3, CaF_2, and ZrO_2, all three of which are notorious carriers of actinide elements. If the clarification process is not totally complete, these materials will ride through extraction processes as interfacial residues[8] and settle out in process fractions, significantly lowering the DF throughout the entire process. Removal of these materials is a difficult and tedious task and must not be underestimated. Such insoluble material should be routed directly to HLW fractions to get it away from subsequent process streams.

The supernatant solution from calcine dissolution will contain large amounts of F, Ca, Al, Zr, Na, B, and Fe ions, many of which interfere with and greatly complicate TRUEX processing. Several of these interfering ions can, and will, precipitate with the purified fractions as they are stripped from the organic stream. Zr and Fe will load the CMPO extractant phase and compete directly with actinide ion extraction; fluoride ions will cause precipitation with the stripped actinide and lanthanide fractions; and Zr will precipitate with phosphate ions resulting from degradation of the 1-hydroxyethane 1, 1-diphosphonic acid (HEDPA) stripping agent. These are serious problems inherent to TRUEX extraction technology and must be addressed during process development. There are potential cures for some of these problems, but they, too, cause additional downstream difficulties. It is necessary to develop the entire process sequence and not simply address specific problems in any single unit operation as they arise.

One approach, in addition to the adaptation of the TRUEX process to a feed stream derived from INEEL calcine, is to consider suitable front-end processes to reduce the amount of key interfering species.[9] One specific problem that a front-end or alternative process could eliminate is the high concentration of species such as fluorides and phosphates. These species cause precipitation of other elements that can mechanically damage the solvent extraction contactors during the back-extraction phase, thereby degrading separations. The committee strongly recommends a careful trade-off study to technically evaluate front-end options.

Current Laboratory Experimentation

INEEL chemistry personnel are aware of some of the problems outlined above, after having performed recent trial extractions on SBW solutions (Law et al., 1996) that produced interstage precipitates and unreasonable process material balances. Further, it appears that much of the analytical data received for the process fractions may be in error by as much as a factor of two (Law et al., 1998: Table 5, page 13). For example, the quoted distribution coef-

---

[8] These interfacial residues are semi-stable emulsions that contain sparingly soluble, colloidal-size particles that ride on liquid interfaces by surface tension effects.
[9] These front-end processes would be used to concentrate actinides apart from other species that could interfere with the TRUEX separations chemistry. An example of such a front-end process of the kind that might be considered would be precipitation of a hydroxide with ammonia, leaving a basic supernate containing elements such as Na, K, Ca, Ba, Sr, and Cs. SLS would remove the precipitate. Boiling off excess ammonia would create an acidic ammonium nitrate solution for Cs ion exchange and SREX. Hydroxide precipitate dissolution in dilute nitric acid followed by fluoride precipitation would isolate the actinide and lanthanide elements as a precipitate. The supernate would contain species such as Al, Zr, and Fe but not Cs, Sr, or TRU elements, and therefore would be a candidate for a low-level waste (LLW) stream. After redissolution of the fluoride precipitate in dilute nitric acid containing borate, the standard TRUEX chemistry then can be applied to extract actinides. Other variations are (1) fluoride precipitation—the addition of fluoride will likely precipitate actinides along with rare earth elements; (2) sulfate precipitation— relatively less information is known about sulfate complexes, formed by the addition of sulfate, but certainly the calcium will precipitate as $CaSO_4$; (3) LaF_3 scavenging (see SBW Option 5 in Chapter 12) could be used on calcine redissolved in fluoride solution to separate a TRU fraction (also containing lanthanides) as a solid precipitate. Most of these processes are based on metathesis, to precipitate highly insoluble actinide and rare earth fluorides from more sparingly soluble species (e.g., CaF_2) in an environment with an excess of fluoride. The committee cautions that these remarks are speculative and offered only as illustrations of the kinds of techniques that could be considered if TRUEX proves to be too complex and costly to implement.

ficients for ^{235}U and ^{238}U (same element) differ by a factor of approximately 3, and for Pu the variation between two isotopes differs by ± 50 percent. Law et al. (1998: Table 4) shows material balances varying from 53 to 130 percent. The committee believes that it is not reasonable to design operational process systems based on so wide a data spread. More detailed and careful development is needed before any actinide separations option can be rationally selected or rejected.

Questionable Experimental Process Modifications

Herbst et al. (1998) describe a process option using a "low-acid" feed (i.e., approximately 1.1 M H+) to minimize extraction of Zr into the CMPO solvent and using a dilute NH_4F scrub solution to remove tramp Zr from the organic phase following the extraction section. While this approach was partially successful in reducing extracted Zr, the follow-on HEDPA strip section apparently precipitated zirconium phosphate. This result indicates that the proposed process is not yet workable. Furthermore, using a low-acid concentration for calcine dissolution will increase the UDS, perhaps significantly. Dilution of standard feed might be a better option, but would dramatically increase process effluent that would then require concentration (presumably by evaporation) to generate an acceptable waste form.

Required Technical Demonstration of Proposed Process

A pilot-scale demonstration project is necessary to show adequate DFs for actinide extraction. This project needs to be performed on an adequate scale using representative samples of aged calcines from all calcine types. These feed samples may be retrieved from the bins by any of several techniques (e.g., core sampling or coaxial suction-lift sampling).

The final demonstration run (a proof test) must use predefined operational parameters and be carried out without parameter changes throughout the run to confirm the suitability of the process to compensate for the known variability in the existing bins. To guarantee the validity of any Class A process, it would be necessary to provide consistent DFs of at least 20,000 for the alpha-emitters to allow for fractional contributions to LLW limits by other species such as Cs, Sr, and Tc. The order of progression of the various unit operations will also be critical since the sequence of steps needed to provide acceptable isolation of Cs, Sr, and TRU will affect each of the chemical operations under consideration.

To reduce technical risk (and potential expense) to a reasonable level, the chemical operations needed to perform TRUEX separations should be developed and demonstrated prior to the commitment of resources to construct full-scale plant facilities designed for this purpose. This demonstration should be funded as a research and development effort prior to any commitment to partition actinides from calcine.

SEPARATIONS PROCESSING CHALLENGES ASSOCIATED WITH THE COMBINATION OF INDIVIDUAL STEPS

Removal of Cs, Sr, and TRU components will require several operations, in addition to the actual separations steps. These operations include retrieval from the tanks or bins, blending, dissolution of most of the solids, solid—liquid separation to provide a highly clarified process feed, preparation of a quantity of suitable and uniform feed for the processes, and finally the separations processes themselves. However, the current database is so limited that there has not been a clear demonstration that some of the methods proposed will, in fact, meet the requirements for the range of feeds that will be encountered. In particular, operating conditions have not been (and cannot be) defined adequately using the limited data that are currently available.

Given adequate process development efforts, the committee believes that separations processes can be made to work, but essential parameters in their application to a full-scale system cannot yet be defined. Examples are the mode of operation of the unit operations, flowsheet integration, equipment design, the actual DFs that will be attained, the amounts of HLW and LLW generated, and operating and capital costs. Moreover, some unit operations may achieve the necessary DF, but only at the expense of generation of excessive HLW (and associated disposal cost). It is counterproductive to spend more money carrying out design studies and cost estimates based on the current state of process understanding, rather than spending it on getting useful calcine characterization data that would reduce the uncertainty associated with the process steps discussed in this chapter. To design a processing plant cogently, DOE must *know* what the starting material is and what requirements the final product must meet. DOE must then develop processes that will address these conditions.

In the present program, some key problem areas have received insufficient attention. Examples are (a) blending to generate a feed composition sufficiently stable for the processes to operate with high performance, (b) methods to compensate for chemical interferences arising from existing elements present in the calcine, (c) methods to deal with all the recycle streams, including spent ion exchange sorbents and extraction solvents, and (d) an effective SLS system.

The evaluation to date with respect to DF requirements is inadequate. Class A limits appear to have been somewhat arbitrarily assigned to the LLW fraction. These limits can be misinterpreted to mean that the particular isotope under discussion must meet those limits. There are two problems with this concept. First, Class A includes the sum of the fractions rule, and isotopes not intended for removal (e.g., Tc) may contribute a significant fraction of that sum. This reduces the limit for those materials that are removed by treatment. Second, processes have variations in performance for many reasons, such as start-up, shutdown, equipment failure, and composition variations. As a result, it is advisable to design for a substantially larger DF than the minimum defined by Class A limits. A prudent practice is to adopt the design basis for each separation step as more than a factor of 10 below the Class A limit.

The technical risk is substantially reduced as the criteria for the LLW fraction are relaxed from those of USNRC Class A limits. Thus, for example, achieving the USNRC Class C limit would require little decontamination except for TRU removal. However, this sort of reasoning should not be used to justify setting an arbitrary limit like Class C as a goal. Instead, the actual criteria must be established in concert with the requirements of the disposal site, an issue taken up further in Chapters 8 and 10. The practicality of more lenient criteria than USNRC Class A limits should be examined from the perspective of both the waste treatment processes and the disposal impacts.

SUMMARY

In summary, there are known treatment methods to accomplish each of the required unit operations to dissolve and process the HLW calcine. The INEEL technical literature shows consideration of many, but not all, reasonable candidates for these treatment methods. Based on the information provided to the committee and the collective experience of its members, the committee believes that it is likely, but not certain, that the objective can be met for each unit operation under well-controlled conditions. It is much less likely that the objective can be met for integrated operations under realistic plant conditions, without encountering undesirably complex operational problems, exorbitant costs, and generation of excessive amounts of secondary wastes. Specifically, the technical risk is substantial if Class A LLW is required. It probably can be done, but at an ultimate cost in both time and money that may be unacceptable. A major development program would be required to reduce the risk and the uncertainty to an acceptable level.

4

Treatment of Sodium-Bearing Liquid Waste

This chapter discusses the sodium-bearing liquid waste currently stored in tanks and managed within the high-level waste (HLW) program at the Idaho National Engineering and Environmental Laboratory (INEEL). This sodium-bearing waste (SBW) is the product of site operations such as decontamination activities, some of which use dilute sodium hydroxide to wash surfaces and solubilize residues. As a result, significant sodium nitrate salts are present in the SBW solutions. For example, waste tank WM-183 contains approximately 4.8 moles of nitrate per liter, in a nitric acid (2.0 M) solution (Law et al., 1997: Table 2, page 5). The cation content, apart from the acid, is 0.7 M Na, 0.6 M Al, 0.1 M K, 0.06 M Fe, 0.04 M Ca, and a small amount of Zr. The radioactive content is about 0.2 curies (Ci) per liter each of ^{90}Sr and ^{137}Cs, and the actinide activity is approximately 500 microcuries per liter (μCi/l), composed of approximately 350 μCi/l from ^{238}Pu and 125 μCi/l from ^{239}Pu.

The relatively high sodium content makes these solutions unsuitable for direct calcination in their present form because sodium nitrate melts at low temperatures and will not produce a granular, free-flowing calcination product. The SBW can be calcined upon addition of aluminum nitrate. This calcine product has been added to the inventory of HLW calcine already in the bins. However, because SBW does not come from spent nuclear fuel (SNF) reprocessing, it is not initially a HLW but rather a mixed low-level transuranic (TRU) waste that becomes HLW by this mixing with a HLW stream. This current management option for SBW, and future treatment options, cannot be considered apart from classification and disposal issues, as described in this chapter.

PRESENT STATUS AND PROGRAM PLANS

The ongoing conversion of low-level SBW to solid HLW by calcination and transfer to HLW storage bins will soon be terminated when the calciner ceases operation on or before June 1, 2000, at least until the calciner is repermitted. The remaining SBW will be consolidated into a few tanks. At this point, the tanks will contain approximately one million gallons of acidic solution and a small but unknown amount of undissolved solids (UDS). This waste is mobile (in liquid form), and it is stored in single-shell tanks located inside concrete vaults. These vaults are not qualified to be an appropriate secondary containment layer under Resource Conservation and Recovery Act (RCRA) land disposal requirements and may not be qualified to withstand seismic activity. In contrast, all of the HLW already has been solidified as calcine and stored in bins that are permitted and designed for a life of 500 years. Therefore, the SBW appears to present a greater current and near-term continuing risk than the calcine, and the committee believes that the solidification and disposition of this liquid waste must be given a correspondingly high priority. The existing regulatory and other legal drivers clearly reflect this priority as well.

PROCESSING METHODS

As one approach to treating SBW, the methods considered in Chapter 3 for processing dissolved HLW (calcine) could be used for SBW with little modification. The solid—liquid separation (SLS) step will probably require different filter media because solids in SBW are certainly different in composition and physical properties than those arising from leached calcine. Other process steps (removal of TRU and possibly Cs and/or Sr, if necessary) could be essentially identical to those for dissolved calcine. However, one possible difference between separations processes designed for SBW and those designed for HLW calcine would be the effects of organic constituents that may be in the SBW. These organics are presumed to be largely decomposed and/or volatilized during calcination and therefore largely absent from the HLW calcine. The solidification process would not necessarily be the same as for calcine because the radioactivity is lower in SBW and the composition is easier to deal with because it is much lower in Zr, F, and Ca than most of the calcine. In contrast to calcine, SBW is more like low-level waste (LLW) at other Department of Energy (DOE) sites where considerable development work has been done, so there is a large base of information regarding waste treatment, vitrification, and grouting. In general, SBW should pose less of a treatment and disposal problem than calcine.

Although many combinations of processing options were presented to the committee, most of these options treat the SBW as if it were HLW, insofar as both SBW and HLW calcine are processed to develop high-activity and low-activity fractions and material waste forms from these fractions. These high- and low-activity waste forms would be sent to different repositories, a future geologic site in the first case and a U.S. Nuclear Regulatory Commission (USNRC)- or DOE-regulated shallow land burial site in the second.

However, SBW is not HLW, but rather is a mixed low-level TRU waste. It is not clear to the committee why this waste is currently being calcined in a manner which converts it to HLW, nor why other options more appropriate for low-level TRU waste were not considered for the disposal of SBW. Examples of these other options are given in Chapter 12. Disposal requirements defined by the repository are one of only a few sources of boundary conditions that establish processing requirements. Since the SBW is not initially HLW, it is counterproductive to convert the SBW to HLW or to force it into a disposal plan designed for HLW.

An additional argument against calcining the SBW is that this calcination involves an addition of aluminum nitrate, thereby increasing the waste mass. If subsequent dissolution of the calcine were then done to separate radioactive species from nonradioactive species, it would seem that calcination (with the addition of aluminum nitrate) would complicate the separations objectives. Instead, a solidification method that does not add chemicals to the SBW would be preferred from this perspective.

REPOSITORY AND TREATMENT OPTIONS FOR NON-HLW

There are two current DOE repositories that are candidate recipients of products of SBW processing. The first is the Waste Isolation Pilot Plant (WIPP) in New Mexico, designated for TRU waste generated in the DOE defense program, but specifically not for HLW. It may be possible that treatment of SBW would lead to a TRU product that could qualify for acceptance at WIPP.

If a separation step were done to isolate a LLW fraction that could be classified as non-TRU and nonmixed, this LLW fraction could potentially go to a DOE LLW disposal facility such as that at the Hanford Reservation or at the Nevada Test Site (NTS). These Hanford and NTS facilities must comply with applicable RCRA regulations, as must WIPP, which is exempt from the RCRA Land Disposal Restrictions (LDR) but not from waste characterization requirements. There are established Waste Acceptance Criteria (WAC) for the WIPP, Hanford, and NTS LLW disposal facilities (DOE, 1996a; Fluor Daniel, Hanford, Inc., 1998[1]; and DOE, 1997). Chapter 9 treats these disposal site options in greater detail.

Other Department of Energy Office of Environmental Management Plans for Disposal of Low-Level TRU Waste in WIPP or NTS

Several DOE sites are planning to prepare low-level TRU waste for shipment to WIPP. Non-TRU LLW is currently being shipped to NTS from several DOE sites. An example from the Oak Ridge National Laboratory (ORNL) is the detailed consideration of several options for treating nearly one million gallons of LLW that is rather similar to the SBW except that it has been neutralized. Some treatment options involve partitioning into TRU and non-TRU fractions, and some involve direct solidification (e.g., evaporation to a saltcake or incorporation in grout). Recently, a privatization contract was awarded to Foster-Wheeler to complete the solidification and disposition of this waste by 2008. The company plans to separate the alkaline-insoluble solids (sludge) from the supernate, yielding a sludge fraction that is a TRU waste and a supernate that is non-TRU LLW. Both will be evaporated to a dry form. The ORNL plans call for sending the sludge solids to WIPP and the supernate fraction to NTS as a dry saltcake (DOE, 1998b; Brass, 1998).

A simple treatment option similar to this should be appropriate for the SBW. This approach would decouple SBW disposition from calcine treatment and is not complicated by disposition of calcine. Further discussion on such treatment options is contained in Chapter 12.

---

[1] This is the Hanford Site Solid Waste Acceptance Criteria for June 29, 1998 from <http://www.hanford.gov/wastemgt/wac/whatsnew>.

5

Vitrification

Vitrification has a long track record as a process for immobilizing nuclear waste and is being used at national facilities such as the Savannah River Site and West Valley Demonstration Project. At the Idaho National Engineering and Environmental Laboratory (INEEL), research on waste vitrification has been conducted since the 1960s. The method consists of mixing the high-level waste (HLW) with glass powders ("frit"), melting the mixture at high temperature (e.g., 1150 °C)[1] in a joule-heated melter lined with ceramic brick,[2] and pouring the melt into waste containers. After cooling, the resulting nuclear waste glass products are generally homogeneous, noncrystalline materials with high chemical durability. They are essentially free from an undissolved crystalline phase but can contain some precipitated crystalline phases and separated amorphous phases as minor phases to the extent that their overall chemical durability is not affected adversely.

This chapter begins by evaluating the vitrification plans for the INEEL HLW and associated developmental studies that were contained in presentations to the committee and in various technical reports cited below. The nonseparation option avoids chemical separation steps by vitrifying the solid calcine directly. The separation option vitrifies the high-activity waste streams resulting from calcine dissolution (including undissolved solids; see Chapter 2) followed by chemical separations processing (see Chapter 3). These two options, and the status of developmental testing conducted to support them, are discussed below in greater detail. Both of these options use continuous, joule-heated melters and borosilicate-based glass compositions.

The chapter ends with a discussion of potential problems for these vitrification approaches, and mentions possible solutions. A fuller treatment of technological alternatives is deferred to Chapter 7, following discussion in Chapters 6-7 of other immobilization approaches. That discussion does not summarize all emerging and viable vitrification techniques such as plasma furnaces or induction-heated cold crucible melters that can be considered for highly refractory materials, but focuses on some developments that in the committee's view were worthy of further consideration for application on INEEL HLW.

---

[1] This temperature limit is used in INEEL technical reports cited in this chapter, as a target operating temperature for a continuous melter.
[2] A technical difficulty lies in vitrifying highly refractory materials (e.g., alumina and zirconia calcines) using a furnace of brick made primarily of alumina and zirconia.

NONSEPARATION OPTION

In the nonseparation option, HLW calcine is directly mixed with glass frit and converted to glass. At present, there are approximately 4,000 m³ of HLW calcine at INEEL, which consists of alumina- and zirconia-based calcine and zirconia-sodium blend calcine (Knecht et al., 1996). In addition, if all existing and projected future liquid wastes are solidified, approximately 2,000 m³ of additional calcine will be produced (Palmer, 1996) primarily from sodium-bearing waste (SBW). In this option, it is estimated that a total of 14,000 "Savannah River-size" canisters (approximately 0.7 m³ each) will be filled with glass with a waste loading of approximately 20 to 30 percent by weight.

Waste Compositions and Characteristics

The base glass composition developed for use with Savannah River and West Valley HLW is a borosilicate type, with the relative abundance of frit constituents tailored to the waste composition to ensure production of a vitrified form with adequate waste loading. However, the calcine compositions at INEEL are quite different from the waste compositions found at these other facilities, in that the INEEL HLW calcines contain significantly higher concentrations of chemicals that have an important effect on glass formation and properties. For example, the alumina-based calcine contains 82 to 95 percent by weight Al_2O_3, the zirconia-based calcine contains 21 to 27 percent by weight ZrO_2 and 50 to 56 percent by weight CaF_2, and the SBW contains high concentrations of sodium nitrate. These waste compositions affect the glass formulation required to achieve a high waste loading. Accordingly, new nuclear waste glasses have to be designed to accommodate these unique compositions. These glass formulation efforts at INEEL are summarized next.

Zirconia-Based Calcine

In the 1970s and 1980s, research was performed at INEEL to develop waste glass compositions for zirconia-based calcine. A borosilicate glass frit "127" with the composition SiO_2 70.3, Na_2O 12.8, Li_2O 6.2, B_2O_3 8.5, CuO 2.1 percent by weight was developed. With 25 to 35 percent by weight calcine and 65-75 percent by weight frit, the mixture was melted at 1100 °C for 3 hours (Staples et al., 1983). Glasses were prepared both on laboratory and pilot-plant scale using simulated nuclear waste calcine. Radioactive waste calcine was also used in laboratory-scale melting. The zirconia content of the resulting glasses was approximately 9 percent by weight (Staples et al., 1983). The durability of these glasses, as tested in the materials characterization center (MCC) tests MCC-1 and MCC-2, was comparable to that of the Savannah River glasses. There was no significant difference in the glass characteristics between samples prepared in laboratory scale and pilot-plant scale as well as between simulated and radioactive calcine.

Alumina-Based Calcine

The glass developed for the zirconia-based calcine cannot be used for the alumina-based calcine nor for a mixture of zirconia-based calcine and alumina-based calcine with more than 15 percent by weight alumina calcine (Brotzman, 1978). Viscosity measurement and Soxhlet leach test of phospho-borosilicate glasses with 23 mole percent (approximately 31 percent by weight) alumina-based waste loading were tested to develop a satisfactory nuclear waste composition that can be melted at 1100°C (Brotzman, 1978). The base composition used was SiO_2 3.2, B_2O_3 31.3, P_2O_5 18.9, Li_2O 4.0, Na_2O 8.3, CuO 3.2, and Al_2O_3 calcine 31.1 percent by weight. A subsequent report (Cole and Colton, 1982) indicated that for alumina-based calcine, the borosilicate glass "532" with the composition SiO_2 46.0, B_2O_3 10.0, Na_2O 7.0, Li_2O 7.0, CuO 2.0, TiO_2 5.0, and alumina-based calcine 23.0 percent by weight, gave the best durability among several tested glass compositions.

SBW to Be Calcined or Directly Immobilized

Glasses were developed (Vinjamuri, 1995) for directly vitrifying the SBW. The frit consists of borosilicate glasses with high (>80 percent by weight) silica content. A glass was prepared with a waste loading of more than 20 percent by weight, and its durability was found to be satisfactory. Specifically, the leach rate normalized to glass composition was low.

Mixtures of Alumina Calcine and SBW

Glasses were developed to vitrify a waste consisting of a mixture of alumina calcine and SBW. In particular, blends of pilot-plant alumina calcine and SBW can be immobilized using the same high-silica borosilicate glass compositions that were developed for the SBW only (Vinjamuri, 1995). For example, a blend of the simulated waste consisting of 57 percent by weight alumina calcine and 43 percent by weight SBW was mixed with a frit consisting of 88 percent by weight SiO_2 and 12 percent by weight B_2O_3 with the waste loading ranging from 20 to 35 percent by weight. The mixtures were melted at 1200 °C or 1600 °C[3] for 2 to 5 hours and their durability measured by the MCC-1 leach test. Glass "532" developed for the alumina-bearing calcine is reported (Cole and Colton, 1982) to be satisfactory for the blend consisting of alumina-based calcine and SBW in the ratio of 2.5:1.

Vitrification Facility and Processing

In addition to the development of adequate glass compositions to accommodate the HLW calcine, other developmental challenges are associated with the design, construction, testing, and operation of a full-scale vitrification plant to operate on radioactive waste feed. INEEL has made detailed plans to build a facility to vitrify calcine waste (Lopez and Kimmitt, 1998). The operation includes blending of the calcine, sampling of the calcine to determine its composition, mixing the calcine with an appropriate glass frit selected from six different compositions, delivering the mixture to the melter, melting and pouring the glass into a metal canister, processing of off-gases, and finally transporting the canister to the interim storage facility. According to this study, the planned facility has a capacity to produce, on average, 3.9 "Savannah River size" canisters per day filled with glass.

---

[3] These temperatures were those reported in INEEL literature for leach testing, and are not intended to simulate continuous melter conditions.

SEPARATION OPTION

In the separation option, calcined waste is dissolved and then separated into high-activity waste (HAW) and low-activity waste (LAW), and the HAW portion is vitrified (Murphy et al., 1995). The incentive for the separation option is the reduction in the vitrified waste volume, which is estimated to be less than one-tenth of that without separation, with the number of canisters expected to be 1,100 instead of 14,000 (Murphy et al., 1995). As a result of the separation process, the HAW will contain a high concentration of zirconia or of potassium and phosphate. For example, the ZrO_2 content in HAW produced from zirconia-based calcine is estimated to be 95 percent by weight, and the K_3PO_4 content in HAW produced from alumina-based calcine and sodium-bearing calcine is expected to be 75 percent by weight and 85 percent by weight, respectively (Staples et al., 1998). These unique waste compositions require special consideration in developing host glasses that can contain reasonable amounts (approximately 25 percent by weight) of these compositionally unusual wastes.

Waste Compositions and Characteristics

As with the nonseparations option, development work has been initiated on glass formulations that would be suitable for the compositions of waste to be vitrified. Because the above-mentioned HAW compositions resulting from the separation process are unconventional, a group effort was instituted to develop appropriate nuclear waste glass compositions (Staples, et al., 1998). The group consisted of personnel from INEEL, Pacific Northwest National Laboratories (PNNL), and the Savannah River Technology Center (SRTC).

This group developed and tested glasses designed to immobilize the compositions of HAW that are estimated to be generated from processing the various types (e.g., zirconia and alumina) of INEEL HLW calcines and SBW. Various borosilicate glass compositions were melted at 1,150 °C for 4 hours and their properties (e.g., viscosity, liquidus temperature, and chemical durability) were evaluated with the objective of producing a nuclear waste glass that can accommodate at least 19 percent by weight of the "all-blend" HAW waste composition. The chemical durability of some of these glass compositions, as determined by the product consistency test (PCT), was found comparable to the Environmental Assessment (EA) glass. Phase separation was observed in the nuclear waste glasses when the phosphate concentration exceeded 5 to 7 percent by weight. In the borosilicate glasses studied, the waste loading was limited by the zirconia and/or phosphate content of the final waste glass composition. Upper limits of approximately 15 percent by weight zirconia and/or 3 to 5 percent by weight phosphate were derived for the final waste glass composition (Staples et al., 1998). This work is still in progress.

Vitrification Facility and Processing

At the present time, there is no detailed plan to build a vitrification facility to process the HAW, but a scaled-down version of the vitrification facility for the nonseparation option presumably could be used.

POTENTIAL PROBLEMS

As noted in Chapter 2, the calcine compositions in the bin sets vary widely, containing large amounts of components like ZrO_2, Al_2O_3, and CaF_2 that are only sparingly soluble in common borosilicate glasses, or components like P_2O_5 that cause phase separation in these same glasses. These calcine compositional variations and chemical constituents pose potentially serious problems for achieving high waste loading in borosilicate glass, as discussed below.

Blending to Achieve Uniform Waste Composition

The retrieval and blending during retrieval (Chapter 2) of calcines of different compositions raises the issue of how to achieve a sufficiently uniform composition of a waste stream for a vitrification process. The waste stream in view here is either undissolved solid calcine, or calcine dissolved in acid solution with some constituents removed. One solution is a design that blends large enough quantities of different types of calcine (after separations steps, if any), by mixing them in a large "feed tank," and to design a frit composition that is compatible with the feed tank calcine composition. A quasi-batch mode would use successive vitrification campaigns to work off successive feed tank volumes; alternatively, if the feed tank were large enough to provide for on-line compositional monitoring of the feed and subsequent adjustment of frit composition, a continuous mode of operation might be possible.

In this approach, an appropriate glass frit (or individual oxide components) can be chosen from among several candidate compositions developed for this purpose. The compositional adjustments to the frit would be minor if the compositional variations of the blended calcine feed were likewise minor. However, the compositional changes in frit could be large if the compositions of the successive blends of calcine are considerably different. Since there are several bins to choose from in recovering the calcine, the compositional variations from blend to blend can be minimized by proper choice of the bins to be emptied. These variations would be prudent to plan for, in the absence of sufficiently large-scale blending to produce a uniform calcine mixture, because of the differences in calcine types stored within each bin.

Waste Loading in Borosilicate Glass

Since alumina and zirconia are only slightly soluble in borosilicate glass at the planned melting temperature of 1150 °C, calcines rich in these oxides would have a low solubility in glass and a relatively lower (probably 15 to 20 percent by weight) waste loading than the 27 percent by weight achieved at Savannah River. If separations alter the content of species such as zirconia, the waste loading would be correspondingly affected.

High Phosphate Content

In the mid-1960s, a phosphate glass was studied for possible use as a nuclear waste glass. Since 1970, however, only borosilicate glasses have been investigated on a large scale in the United States, with two exceptions: phospho-borosilicate glass compositions for a separation option (Staples et al., 1998) and lead-iron phosphate glass development (Sales and Boatner, 1988). In view of the unique composition of the various wastes at INEEL, especially those of high P_2O_5 concentration, new phosphate glass compositions deserve attention (Day et al., 1998), because such phosphate glasses could contain larger amounts of these unique wastes and have equally good chemical durability. While borosilicate-based nuclear waste glasses appear to exhibit phase separation that leads to chemical durability deterioration when the P_2O_5 content exceeds 5 to 7 percent by weight (Staples et al., 1998), depending on the final waste form composition, phosphate-based glass should be able to avoid this problem. Therefore, further examination of phosphate-based compositions would be useful to probe issues of chemical durability and corrosion.

Zirconia-Related Problems

Zirconia has a limited solubility in oxide glasses and slow dissolution kinetics, especially at the melting temperature of 1150 °C proposed for nuclear waste vitrification. Therefore, process options other than vitrification may be more expedient ways to handle zirconia-rich waste. In the production of a homogeneous vitrified waste form, the zirconia would be dissolved completely. However, zirconia may be more useful in a crystalline form serving as a host for some radioactive elements. Indeed, there is a growing literature (e.g., Heimann and Vandergraaf, 1988; Oversby et al., 1997) on the use of zirconia as an inert fuel matrix and durable waste form. A glass/ceramic, made by melting or sintering, may be a better waste form for zirconia-rich materials than a vitrified waste form, since a higher waste loading is expected (resulting in fewer canisters), and it should have an acceptable chemical durability.

Tolerance of Glass to the Content of Calcium Fluoride

The calcium fluoride component of the feed to a vitrification process is potentially problematic. The concern regarding fluorides such as calcium fluoride stems from their tendency to vaporize at high temperatures (which results in the possible formation of corrosive off-gas constituents containing fluorine), their limited solubility[4] in glasses such as borosilicates, and their tendency to promote phase separation (which has the potential to reduce chemical durability or increase the rate of refractory or electrode corrosion) in borosilicate glasses. These potential problems can be avoided by efforts such as (1) removing the fluorides such as calcium fluoride from the waste prior to vitrification,[5] (2) designing the glass composition to accept a small amount of fluoride content,[6] or (3) using glasses other than borosilicates that are more compatible with fluorides. Fluorides are typically more soluble in phosphate glasses than in borosilicate glasses; as an example, fluorophosphate glasses are well known in the optical glass field and have been manufactured commercially for several decades.

---

[4] Although most of the fluoride evaporates into the off-gas, some remains to enter the glass medium.
[5] This entails consideration of the undissolved solids (UDS), which is likely to contain the stable fluoride CaF_2, among other ingredients, and which is likely to be sent to the high-activity fraction for immobilization.
[6] In view is a fluoride content less than one weight percent. Slightly higher contents would probably cause precipitation of insoluble fluorides during cooling, such as Na_3AlF_6 (cryolite) or Na_2SiF_6, but this may be acceptable in a waste form.

Testing Needs

If Joule-heated melting is chosen for the INEEL HLW calcine, prototype testing will be needed. Important issues that should be addressed would include corrosion of the refractories and electrode materials, atmosphere and off-gas controls,[7] and facilities for draining the melt into canisters and/or for removing insoluble materials (e.g., noble metals) from the furnace floor. In the absence of waste composition data, these required tests cannot be specified more definitively. A minimum requirement, however, should be that corrosion tests with candidate refractories and electrode materials be performed using simulated, if not actual, calcine waste, to exhibit behaviors important to the design of a full-scale process.[8]

---

[7] Losses of volatile elements occur with any high-temperature process, and are well known for both phosphate glasses and borosilicates (e.g., with the latter, these losses are thought to be due to either compound formation or entrainment with volatile B_2O_3). The choice of glass composition and process conditions will affect volatilization losses during melting. Further specification of important volatile losses will depend upon knowledge of the calcine compositions and process conditions. These volatile losses can be reduced in melter designs that use a "cold cap" of (unmelted) batch materials, as compared to melter designs such as in-can or in-crucible melting that cannot use cold caps.

[8] In experience to date, although some phosphate glasses exhibit higher corrosion, iron-based phosphate glasses do not corrode common glass contact refractories any more than the Savannah River borosilicate glasses (Chen and Day, 1999; Day et al., 1999); hence both phosphate and borosilicate glass compositions are potentially viable options, as discussed in this chapter.

6

Cementation

Cement-based mixtures have a long history of use for immobilization of low-level radioactive wastes (Moore et al., 1975; Moore, 1981). The conventional immobilization process incorporates waste in a mixture of portland cement, clays, bases, and water. Upon drying or thermal treatment, mineral phases such as feldspathoids, zeolites, and other hydroceramics are formed that contain the waste ions. The waste form is durable and resistant to leaching by water.

The procedures and equipment to produce cementitious waste forms are modifications of practices used to produce ordinary concrete. Mixers, forms (for shaping), and drying and curing equipment would be used in remote or shielded operations.

Concretes, such as those "formed under elevated temperature and pressure" (FUETAP) that were developed at the Oak Ridge National Laboratory (McDaniel and Delzer, 1988), also have been proposed for the immobilization of high-level wastes (HLW). FUETAP utilizes the thermal output of the waste to accelerate the curing process. Development work at the Idaho National Engineering and Environmental Laboratory (INEEL) has built on the FUETAP experience in the study of cementitious waste forms for the immobilization of INEEL HLW (Russell et al., 1998; Dafoe and Losinski, 1998; Lee and Taylor, 1998). These forms have been considered for use on the following wastes:

- the high activity waste (HAW) generated after several separations steps have been performed on the dissolved calcine or sodium-bearing waste (SBW),
- the HAW generated after only one transuranic (TRU) separations step has been performed on the waste stream, and
- the HLW with no separations steps at all; that is, the solid calcine and liquid SBW.

Each of these separations-based and nonseparations-based options have different processing parameters, give different end products, and require different ultimate disposal options. For example, some of these options generate a low-activity waste stream, which is also immobilized in a cementitious waste form.

A technical assessment of the various cementitious processes and end products with respect to the given statement of task was performed and is described in the paragraphs that follow. The cementation processes, described first, are relatively straightforward operations as compared to other immobilization techniques in Chapters 5 and 7. The quantity of final waste that is acceptable to produce and the qualification of the cementitious waste form for suitable disposal are the major issues that would need to be resolved to make cementation a fully viable option. The waste form qualification issues are treated at the end of this chapter, and the issue of the quantity of high-level waste is taken up in Chapters 9 and 10.

PROCESS DESCRIPTIONS

Brief process descriptions of the direct (nonseparations) and separations cases for the waste materials presently at INEEL are given below. Two direct and three separations-based cementation processes have been considered.

Direct Cementation:
Mixed HLW Hydroceramic with Feed of Clay, Slag, Soda, and Water

In this option, the existing SBW liquids would be calcined and added to the other HLW calcine presently in the Calcined Solids Storage Facility (CSSF) at INEEL. This mixed calcine would then be mixed with clay, blast furnace slag, caustic soda, and water such that analogs of naturally occurring feldspathoids/zeolites are generated. This cement/waste mixture would be extruded into stainless steel canisters where a curing process would produce a structurally sound and geologically stable hydroceramic waste product. The canisters used would be the same as those used at the Defense Waste Processing Facility at the Savannah River Site (Dafoe and Losinski, 1998). This process does not minimize the volume of waste to be disposed of and relies on future acceptance of the mixed HLW form (i.e., containing both hazardous chemicals and radionuclides) at a currently unapproved site such as the Greater Confinement Disposal Facility (GCDF)[1] at the Nevada test site (e.g., NTS) (see Chapter 9). That is, this option does not involve the site of the geologic repository for spent nuclear fuel and co-disposed defense HLW. The calcining and cementation would be accomplished within a 20-year operating cycle (Dafoe and Losinski, 1998).

The process outlines have been given (Dafoe and Losinski, 1998) and constitute straightforward processes and process equipment. This process could be implemented simply, with little development, and it has a low risk of failure (Dafoe and Losinski, 1998) and minimum personnel exposure. As a result, costs should be relatively low compared to processes requiring redissolution of existing calcine and various separations processes. It should be possible, due to process simplicity, to shorten the stated 20-year operating cycle quoted in Dafoe and Losinski (1998).

Direct Cementation:
Mixed HLW Hydroceramic with Feed of Sucrose, Clay, NaOH, and Water

The second case of direct cementation is described by Lee and Taylor (1988) and involves different feed materials than the first case described above. In this case, SBW liquids would be slurried with calcine from the CSSF and sucrose and recalcined. This would require the existing calciner to be repermitted. This new calcine would be classified as mixed HLW. The new calcine would then be mixed with calcined kaolinite clay, sodium hydroxide, and

---

[1] The GCDF is a series of boreholes in arid alluvium of the Nevada Test Site that have been used in the past to store Department of Energy (DOE) classified TRU waste (see Chapter 9).

water. The cement mixture would be placed into HLW canisters, steam cured via autoclave, dewatered, degassed, and sealed. These would then be placed in interim storage for transfer to a packaging facility and transport to the GCDF (Lee and Taylor, 1988). The cementitious waste form produced is a hydroceramic designed to be stable in the NTS alluvium. This process does not minimize waste volumes but does use simple and inexpensive feed materials. The process and process equipment are simple, easy to operate and commonly available. The process does rely on future acceptance of a mixed HLW waste form at an as yet unapproved site (e.g., NTS). The calcine and SBW can be processed using this approach in about 5 years (Lee and Taylor, 1988).

The process outlines for this option have been given (Lee and Taylor, 1988) and are again straightforward. Because of process simplicity and simple equipment, the risk of failure is low and personnel exposure should be low. As a result, costs should be low relative to those processes requiring calcine redissolution and various separations. This option does depend on the existing or similar calciner being usable soon and probably entails the lowest risk and lowest development effort of any of the various cementitious processes.

In both of the two direct cementation processes described above, the resolution of the treatment and disposal of all Resource Conservation and Recovery Act (RCRA) hazardous materials must be pursued energetically because it may be difficult to dispose of mixed HLW. In addition to the RCRA problem, there would be approximately 13,000 cubic meters of cemented waste formed with the first process (Dafoe and Losinski, 1998) and approximately 12,000 cubic meters with the second process (Lee and Taylor, 1988). Although neither of these volumes should be a problem if the GCDF is made operable, the committee believes any planned use of a GCDF must recognize that the use of such a disposal site for INEEL HLW is in a very early stage of conceptual development and outside the current regulatory approach (see Chapter 9).

Separations-Based Cementation

Three processing options, briefly discussed below, involve separating specific components from the HLW and SBW to decrease the amount of HLW requiring disposal at a HLW repository. The bulk of the waste volume remaining after these separations are performed would be grouted into a low-level waste (LLW) form.

TRU Separations with Class C grout

In this option, TRU components are separated for disposal at the Waste Isolation Pilot Plant (WIPP) and the resulting LLW waste would be grouted for disposal as Class C waste (Option 2.2.2 of Russell et al., 1998). Further details are provided in Landman and Barnes (1998).

The TRU separation processes have only undergone small-scale demonstrations at INEEL. The major advantages of this option are (1) the use of only TRU separations (i.e., Cs and Sr separations are avoided), (2) a single off-gas treatment facility, and (3) the WIPP repository, which is potentially available in the future to accept mixed TRU waste. A regulatory reclassification ruling would be required because this option proposes to produce no HLW outputs.

HAW and TRU Separations with Class A Grout

This option involves separation of two fractions—HAW and TRU—from the LLW stream (Option 2.2.3 of Russell et al., 1998). The HAW fraction, containing both cesium and strontium, would be processed into a vitrified HLW product. As in the previous option, the separated TRU fraction would be dried and packaged for disposal at WIPP. After separations, the LLW would be concentrated by evaporation, denitrated, and grouted as a Class A waste.

This option requires significant dissolution and separations steps that have not been demonstrated in pilot-plant level operations. A regulatory reclassification ruling would be required, as above, to approve of a TRU waste stream derived from a HLW input.

Separate TRU, Cs, and Sr Separations with Class A Grout

The third separations option (Option 2.2.1 of Russell et al., 1998) separates three fractions—TRU, Cs, and Sr—from the LLW stream. These three separately generated fractions are then combined and vitrified into a HLW product. The remaining LLW would be grouted after denitration for disposal as a Class A waste. Again, further details are provided in Landman and Barnes (1998).

The processing disadvantages and regulatory issues are similar to those of the previous two separation processes. Because Class A grout has fewer disposal restrictions than Class C grout, the Class A LLW grout of this and the previous option are relatively easier to dispose of, although the RCRA constituents of the grout likely represent a more important restriction that all options need to address. In these last two options, an interim-storage facility would have to be built at the INEEL to store the vitrified HLW until a repository became available for its disposal.

Disposal of LLW Grout

The three separations-based operations all produce a large volume (approximately 22,000 to 27,000 m^3) of LLW grout to be disposed. For this LLW grout, three specific disposal concepts have been considered (Russell et al., 1998). In one, grouted LLW would be pumped into the empty storage tanks and the empty bin sets. In the second, grouted LLW would be packaged in thick-walled, cubic-meter-size concrete containers suitable for disposal in a near-surface facility. In the third, grouted LLW would be packaged and sent to the Hanford site in the State of Washington for disposal.

While the use of the liquid storage tanks and calcine bins for storage of a large volume of cementitious LLW is technically and economically attractive, all of these concepts depend on delisting some of the RCRA wastes (see Chapter 9).

Because of the relative simplicity of cementation equipment and processes, these options for immobilizing INEEL HLW are attractive. No high temperatures are required, and in some options, no chemical separations are required. These conditions lessen hazards to workers and the public. However, because of the many possible mineral compounds formed, depending on the starting materials and the waste constituents, significant further testing is required. This issue is taken up in the next section.

CEMENTITIOUS WASTE FORMS

The cementitious waste form proposed for use on INEEL HLW is described in a recent paper by Siemer et al. (1998). The waste form is not the conventional portland cement-based waste form, but rather is of a special composition based on the alkaline activation of pozzolanic aluminosilicates. In this waste form, aluminosilicates, such as the zeolites (Siemer et al., 1998), are important radionuclide-bearing phases.

The approach builds on previous experience with FUETAP and uses a "mild" hydrothermal processing (identified as a "clay reaction process"). The waste form was prepared by having kaolintic clay react with a range of simulants for the SBW. Samples were cured at approximately 200 °C for several hours. Samples were characterized by x-ray diffraction (XRD) and scanning electron microscopy (SEM), and three types of leach tests [the standard leach tests included the ANSI-16.1 standard, the Product Consistency Test (PCT), and the Toxic Characteristic Leaching Procedure (TCLP)] were performed on the product. Principal phases resulting from the synthesis included unreacted quartz, a hydroxysodalite, and zeolite-A ($Na_{12}Al_{12}Si_{12}O_{48}$ $27H_2O$). The results of the leach tests are interpreted to show that the cementitious waste form provides equivalent or better performance than typical waste form glasses (despite a much higher surface area for the cementitious waste form).

No specific data were presented for the waste form products that would result from the two different processing options of direct cementation and separations-based cementation. It is clear from the literature on cement that a wide variety of phases may form depending on initial compositions and processing parameters.

Despite the considerable literature on cementitious waste forms and the fact that cementation is a candidate technology for INEEL waste, it does not appear that much recent work has been focused on the development and characterization of the cementitious waste form. Indeed, the justification for the use of the cementitious waste form, aside from the ease of processing, is the following: (1) it is comparable to glass in its performance and/or (2) the waste form is of little consequence in the overall performance assessment of a geologic repository. Setting these issues aside, however, there is another major issue, with a number of sub-issues, that should be addressed if the cementitious waste form is to be thoughtfully compared to the alternatives (e.g., vitrification, glass ceramic, sintered glass, or calcine).

The major issue is that there are inadequate data to support the contention that cementitious waste forms can be made more quickly, more cheaply, more simply, and more safely than other (e.g., vitrified) waste forms. That contention may be true, but it should be established by thoughtfully designed experimental programs and an analysis of comparable data (a full-scale demonstration of the process is not required). Such experimental programs and comparable data analyses would include the following:

1. It would be necessary to develop laboratory scale synthesis procedures that reflect both the compositions of materials that will actually be used and the process technologies that will be developed and compared. There is a surprising lack of INEEL laboratory data on the potential synthesis processes and the final products. Systematic studies should be completed to investigate the effects of variations in the composition of the waste stream, variations in waste loading, and the consequent effects on product consistency.

2. A complete and precise characterization of the cementitious waste forms would have to be done. The present use at INEEL of XRD and SEM, as illustrated in Siemer et al. (1998), is cursory at best. Phase identification is tentative, and neither XRD nor SEM will detect many minor but important radionuclide-bearing phases. The characterization must be completed as a function of the initial starting material compositions and over the range of synthesis conditions (e.g., variations in temperature and curing time). It is important that standard, well-described procedures lead to a consistent product. It is also important to establish

the effect of variations in SBW and calcine compositions on the final product. The characterization must include not only the identification of phases, but also the descriptions of the microstructure and surface area and the partitioning of radionuclides into particular phases. This is exactly the information that will be required for any substantive assessment of waste form performance as a function of leaching under different geochemical conditions or in a radiation field.

3. A more fundamental understanding of leaching mechanisms is needed. The three leach tests (ANSI-16.1, PCT, and TCLP) used in previous studies provide only a qualitative basis for comparison of waste form performance. Longer term tests with more complete analysis of the solution chemistry, followed by solid-state characterization of the waste form after leaching, are required. The release, transport, and precipitation of radionuclides may occur over very short distances and result in important changes to the material form that could affect long-term performance. It is also important to establish the stability of zeolitic phases in the cementitious waste forms and in the proposed repository environment.

4. In addition to leach tests to measure durability, it would be necessary to determine the effect of the high level of dissolved salts that accompany radioactive liquid wastes on the durability of the waste form. Recent work (Guerrero et al., 1998) has demonstrated that there can be important effects related to the formation and evolution of reaction products (e.g., brushite, a hydrated calcium phosphate). The formation and dissolution of these reaction product phases can affect the porosity and surface area of the cementitious waste form and the subsequent release of radionuclides.

5. Additional study of the effect of radiolysis of structural water in the hydrated phases in the cementitious waste form is necessary. Radiolysis can lead to the formation of gases and subsequently affect the porosity and mechanical strength of the waste form. This is a phenomenon that has received only limited attention; however, the solid-state radiolysis of compounds that contain structural water may seriously affect the long-term durability of a cementitious waste form.[2]

6. Also required are additional data on the effect of mercury, an important component of some of the wastes. There must be an investigation of the fate of Hg during processing and in the final waste form in order to predict the amount, chemical form, and long-term behavior of any Hg constituents.

SUMMARY

As can be seen from the foregoing, the proposal to use a cementitious waste form, particularly one in which zeolites are used to advantage to retain critical radionuclides (e.g., Cs) may have merit. A principal advantage is that the SBW and/or the calcined waste might be directly incorporated into the cementitious waste form with minimal or no treatment. However, an informed decision on the use of a cementitious waste form must be based on a thorough analysis of experimental data gathered on waste forms typical of the selected processing technology. At present, these data do not appear to be available. In addition, the analysis and presentation of the data that are available are not in a form that allows for an appropriate analysis.

---

[2] In the event that gas generation (through radiolysis or other processes) were to be a problem, the waste form containers would have to be engineered to prevent overpressurization, such as by using recombiner catalysts, which was done in shipments of Three Mile Island core debris to INEEL.

7

Other Waste Forms

This chapter contains general descriptions of several procedures or processes for producing waste forms from the Idaho National Engineering and Environmental Laboratory (INEEL) calcine other than the vitrified (see Chapter 5) and cementitious (see Chapter 6) waste forms already discussed. Except for the hot isostatic press (HIP) option, these processes come from sources other than the Department of Energy (DOE) literature that was presented to the committee. In most cases, these procedures are at an early stage of conception and will require a significant amount of additional investigation to assess their feasibility for practical use with the compositionally unique INEEL calcines. These descriptions are, therefore, necessarily general and are included primarily for purposes of completeness.

Some of these processes, such as those involving sintered glass (Gahlert and Ondracek, 1988) and glass-ceramics (Hayward, 1988), are not new, having been first suggested in the early 1980s. Unlike fully vitrified forms, which are composed ideally of a single-phase, chemically homogeneous glass, the glass/ceramic waste forms are composed of a mixture of one or more crystalline phases bonded together, and the partially vitrified waste forms surround these crystals with a glassy phase.

Most of the waste forms discussed in this chapter will be physically and chemically heterogeneous on a microscale, so that the overall properties such as leach resistance will depend on the chemical properties of each crystalline and glassy phase. However, there are no scientific reasons preventing a properly prepared waste form of this heterogeneous character from having a chemical durability equivalent or superior to that of a fully vitrified waste form. In certain cases, there are important manufacturing advantages in producing heterogeneous, rather than homogeneous, waste forms.

A feature all of the following procedures have in common is that they utilize the existing INEEL calcine in its present state. No dissolution of the calcine or separation of the radioactive elements from the calcine is anticipated. These procedures fall into two categories identified below as simulated spent fuel (SSF) and partial vitrification.

SIMULATED SPENT FUEL

The simulated spend fuel (SSF) process consists of forming the existing alumina and zirconia calcine into strong, cylindrical pellets that are then packed into a suitable metal tube such as zircaloy or stainless steel and sealed (basically the way that ordinary light water reactor

(LWR) fuel rods have been prepared for decades). The final step is to bundle the tubes and seal them in a second (metal) protective canister for final storage. The premise of this procedure is that if the Calcined Solids Storage Facility (CSSF) calcine is packaged to satisfy the acceptance requirements for commercial spent nuclear fuel (SNF) and can be made to emulate the internal properties and containment features of SNF, then the regulatory issues for permanent disposal should be greatly simplified. This premise is unverified at this time, and, of course, Resource Conservation and Recovery Act (RCRA) issues would still have to be addressed.

Much of the equipment already exists for forming pellets approximately 2.5 cm (1 inch) in diameter and height by pressing the existing calcine, sintering the pellets at elevated temperature to increase their strength and chemical durability, and packing and sealing the pellets into metal tubing. These procedures are similar to those that have been used for many years to manufacture commercial fuel rods. These processes would have to be modified and optimized, however, to the INEEL powdered calcine mixture, which might also require additives to improve its sintering properties, strength, and chemical properties. There is very little information available at this time as to what cold pressing (pressure and time) and sintering (time/temperature profile) conditions will be needed for the existing calcine, along with any additives that might be needed, to produce a high-strength, chemically durable pellet with the desired properties.

Another issue for which information would have to be developed is that of the calcine blending requirements. The degree of calcine homogeneity required for this physical pressing process is likely to be different than for a chemical process using materials in solution. Investigations would be needed to address what level of inhomogeneity the SSF process could tolerate, whether blending different ages and types of calcine is an adequate strategy, and whether blending during retrieval would pose any problems.

There are no readily identifiable obstacles to producing a metal-clad pellet from the existing calcine, but considerable developmental work is still needed to determine if pellets of sintered calcine, with acceptable properties, can be produced by this procedure. In those calcines containing fluorides and nitrates, the behavior of these materials during the pressing and sintering steps will need to be determined, as would any tendency of the compacted pellets to release gases or to corrode the metal container.[1]

There is no reason why the SSF pellet should have the exact dimensions of actual fuel. The diameter of a nuclear fuel element is set by heat transfer considerations that would not be the same for a SSF waste form. SSF rods could be considerably greater in diameter than actual fuel rods, thus reducing the number required. Or, they could be packaged in odd shapes, to fill void volumes in a repository.

---

[1] Testing for these behaviors undoubtedly would be required to certify the waste form, but reasons can be advanced for expecting them to be a minor issue. Although NRC (1997) reported on radiolytic degradation of fluorides in a fluoride salt, with generation of fluorine gas and possible corrosion products, the chemical environment in the INEEL high-level waste calcine differs from the environment of the Molten Salt Reactor Experiment (MSRE) fuel salts in its thermodynamic activity for fluorine and oxygen, and hence in the expected reactivity and recombination of fluoride compounds. Because the calcium fluoride (CaF_2) salt in the calcine is structurally stable, the fluorite crystal lattice would tolerate defects. Although radiolysis can break the ionic bonds to create fluoride ions in the crystal, these stripped ions can readily drift to a vacancy, crystal defect, or interstitial site and recombine. The effectiveness of this recombination, which prevents the release of gases such as fluorine, is enhanced by the ionic character of the calcium fluoride bonds. The high electronegativity of fluorine atoms, the lack of oxidizable or volatilizable ions, and the well-known retention of atomic species in strong crystalline matrices assures that not much (likely none) of the elemental fluorine or volatile fluorides would be released. This contrasts with the behavior of the fluoride compounds in the MSRE fuel salts, which have bonds of more covalent character and in which less effective recombination would be expected. There, oxidizable UF_4 was present and formed a volatile product (UF_6), and a mechanism existed for its migration and reconcentration outside the salts. These MSRE conditions are not analogous to those of the SSF waste form considered here.

A potential advantage of SSF is that it does not add any more inert material to the calcined waste. This minimizes waste volume, and also makes this process simpler than the alternatives. This simplicity should translate into lower cost and radiation exposure. Another potential advantage is that transportation issues might be simpler, insofar as the safe transport of properly packaged spent fuel has already been established.

Radioisotopes are more easily leached from calcine than from some other waste forms. An unknown issue is the leach rate of radioisotopes from SSF, and whether this leach rate is acceptable. For example, a high leach rate of fission products such as Cs and Sr may not be relevant to repository performance because these radioisotopes will likely have been diluted and suffered natural decay by the time they would travel to the accessible environment. The leach rates of longer-lived radioisotopes would be more important limitations. Another unknown is the degree to which the waste form would exclude moisture.

Fabrication Issues

SSF pressing and sintering operations for calcine would have similarities and differences compared to these operations for fresh nuclear fuel (UO_2) fabrication. Some of these comparisons are noted below.

1. The firing temperature and redox conditions would differ. Conditions for UO_2 are 1650-1700 °C in a hydrogen atmosphere, with controls on dimensions, density, and stoichiometry. In contrast, the calcine can be heated in oxidizing conditions (e.g., air) since the heavy element valence states are stable. The sintering temperature would be approximately 1000 °C. Residual uranium will be in the high oxidation states, which provide the highest affinity for cesium retention. Silica could be added as a sintering aid to promote bonding and to capture cesium. A variety of commercial furnace systems are available for these conditions and can be utilized relatively inexpensively.

2. Shielded cell and remote-handled operations would be required due to the radioactivity levels of the calcine. Remote-handled technology is already in use in UO_2 and some PuO_2 fuel fabrication plants. Automated powder pressing, sintering, rod loading, and bundle fabrication are also already used. The activity of the calcine implies that these operations be done in a shielded cell, with possible complexities. Nevertheless, such operations would probably impose fewer facility requirements and less cost than chemical processing.

3. Thermal management issues would have to be investigated for SSF production and storage. Although calcine temperatures in the bins are quoted in this report as 190 °C for the zirconia calcine and 440 °C for the alumina calcine, the temperature elevation in a smaller size sample during its production and storage could be controlled to be smaller, as with air streams and cooling channels that can be incorporated in equipment design features to effect adequate heat transfer. Were this temperature rise to be controllable to only a few degrees, heat generation would not be a dominant issue affecting the process design. High temperatures could pose difficulties in blending suitable binders and/or plasticizers into the feed material for pressing.

4. Operational issues in blending and combining with material additives would have to be developed for SSF, but again, fuel fabrication techniques exist as a useful reference to guide this development. Agglomerates of UO_2 and PuO_2 used in fuel manufacturing have sizes similar to the calcine particles. Organic binders are commonly used for pressing and are readily burned away in sintering. Zoned furnace heating or multi-staged treatments at controlled temperatures could be used to provide desired waste form properties.

5. Off-gas treatment would differ due to the presence of fission products and other volatile species. In the SSF concept, fluorides, ruthenium, iodine, and mercury are among those species that should be planned for off-gas entrapment. Since the oxidative sintering of calcine does not require high gas flows, as is required for UO_2 sintering, the off-gas control could be simpler.

In summary, the committee believes that fuel fabrication techniques can be appropriated and modified to develop a suitable SSF process, although significant development and testing would be required. The overall merit of a SSF approach would depend on the qualification and regulatory approval of the final waste form, as discussed next.

SSF Waste Form

The anticipated but unverified advantage of this SSF waste form is an expected ease in its qualification compared with other types of waste forms, insofar as it would be analogous to an already qualified waste form (i.e., commercial spent fuel, which must qualify technically if direct disposal is pursued).[2] Other significant advantages are the large waste loading (which could be up to 70 to 90 weight percent depending upon the additives needed), the availability of existing commercial equipment, the extensive experience in fabricating commercial nuclear fuel by this procedure, an expected high throughput (several presses can be used simultaneously), and the double encapsulation in corrosion resistant materials whose performance is well known. All of these factors should contribute to a relatively inexpensive waste form (as compared, for example, with complete vitrification of the existing calcine). These potential advantages make the SSF concept an attractive one for further study at this stage of development of the INEEL HLW program. However, as stated above, the overall merit of a SSF approach would depend on the qualification and regulatory approval of the final waste form.

PARTIAL VITRIFICATION

A common feature of the following five partial vitrification processes is that the *existing* calcine, plus additives, will be only partially vitrified so that the final waste form consists of crystalline material(s) embedded in a glassy matrix. This type of partially vitrified solid is loosely referred to as a "glass-ceramic," but a true glass-ceramic is a material that at one time was totally glassy and later partially or fully crystallized in a controlled manner.

Unlike full vitrification, where the final waste form is composed ideally of a chemically homogeneous glass free of crystalline particles, these processes yield a chemically inhomogeneous waste form that is only partially glassy. There are no scientific reasons why a partially vitrified waste form cannot be made that will meet all of the existing performance requirements for fully vitrified waste forms. However, at this time, the absence of current certification for partially vitrified waste forms is a perceived disadvantage for partial vitrification.

A common feature of all five procedures listed below is that, as with the vitrification options discussed in Chapter 5, they all use the same type of feed stock, namely, a mixture of

---

[2] That is, a SSF waste form would be analogous to spent fuel material forms primarily in the way it would be produced (as a sintered oxide), and not necessarily in its detailed material form, composition, physical properties, and chemical properties, all of which would be different. Although the radionuclide-bearing phases might differ from those of SNF forms, a SSF waste form might provide sufficient similarity to SNF waste forms that are to be disposed of in a geologic repository, particularly if production similarities were to translate into similarities in material form properties to permit a straightforward comparison of waste form performance.

the existing and future calcines at INEEL. Thus, the first step will be to recover the calcine and blend it together, with or without additives, to form a sufficiently large supply (e.g., enough to last 1 to 2 years) of a mixture of calcine sufficiently homogeneous to bound its important physical and chemical characteristics within an acceptable range.

Cold Pressing Followed by Sintering

In the cold pressing followed by sintering (CPS) procedure, a mixture of calcine and additives (glass formers) is cold-pressed into a shape of the desired dimensions. The cold-pressed, partially porous compact is then heated through a controlled temperature/time profile to form (sinter) a dense, mechanically strong, and chemically durable solid. During the heat treatment, numerous chemical reactions occur between the calcine particles and any additives to form a heterogeneous composite of crystals embedded in a glassy matrix. The purpose of the additives is to control the type(s) of crystal(s) present and the properties of the glassy matrix. The objective is to form a partially vitrified waste form composed of crystals and glass, both of which have a chemical durability that meets existing requirements.

The end result of the SSF and CPS procedures is basically the same—namely, a partially vitrified waste form that should be dense, strong, and chemically durable. However, the size and shape of the waste form would probably be different, and no encapsulation of the waste form in protective metal tubing is envisioned for the CPS procedure.

Three advantages of the CPS procedure are that (1) no further processing of the existing calcine, apart from physical blending operations to achieve sufficient homogeneity, is necessary; (2) the expected high waste loading (greater than 50 percent by weight in the final waste form) reduces the volume of waste to be permanently stored; and (3) the procedure is based on well-established technology for pressing and firing consolidated powders. All of these factors should reduce the cost of a waste form made by CPS compared to complete vitrification.

Hot Uniaxial Pressing

Hot pressing differs from the CPS procedures in that the pressing and firing (sintering) steps are combined into a single step, so that the mixture of calcine and any additives is consolidated under unidirectional pressure in a heated mold. This process again produces a partially vitrified waste form composed of crystals embedded in a glassy matrix.

The hot pressing process has essentially the same advantages and disadvantages as noted for the CPS process. In general, hot pressing is more difficult and requires more complex equipment than cold pressing followed by firing (sintering) in a separate furnace. There are reports of previous work in Germany (Gahlert and Ondracek, 1988) where waste forms up to 30 cm in diameter have been successfully prepared using hot pressing.

In a presentation to the committee, Professor Werner Lutze of the University of New Mexico presented encouraging data from his experiments on a simulant of INEEL zirconia calcine (a surrogate representing the calcine type in CSSF #5) and on a simulated Hanford tank waste form (Lutze, 1998). In these experiments, hot pressing a mixture of the nonradioactive calcine surrogate (30 and 40 percent by weight) with silica and soda produced partially vitrified products. The partially vitrified simulated waste form he produced from the simulated CSSF #5 calcine contained zirconia (ZrO_2) and fluorite (CaF_2) crystals embedded in a chemically durable sodium-zirconia-silicate glass matrix. Table 7.1 summarizes the properties of this partially vitrified "sintered glass" waste form and compares them to properties reported in INEEL literature for partially vitrified waste forms produced and studied in the past at INEEL. Although he used hot isostatic pressing (HIP) to produce partially vitrified waste forms of good chemical durability, he believed that the same results could be obtained for partially vitrified waste forms made by unidirectional hot pressing (i.e., hot uniaxial pressing, or HUP), which is simpler and less hazardous than HIP, discussed next.

Hot Isostatic Pressing

INEEL has HIP experience (Russell and Taylor, 1998) and has produced many simulated waste forms (Staples et al., 1997) by this procedure, starting in the 1970s. The process consists of placing the starting materials (calcine plus any additives) in a suitable (usually metal) container that can withstand the temperature and pressure to be used, and then heating the contents in a pressurized chamber to a desired temperature for the necessary time. The chamber is then depressurized, and perhaps cooled, whereupon the reacted material is removed and the procedure repeated.

There are no scientific reasons why a partially vitrified waste form with the required properties cannot be produced by the HIP process. However, the committee does not consider the HIP process to be a practical method for consolidating the INEEL calcine for three reasons:

1. First and most important are the safety aspects of the HIP process, which uses a highly compressed gas. Because of the high temperatures needed to react and densify the INEEL calcine (at least 1000 °C and probably higher), some type of pressurized gas must be used instead of a less dangerous liquid. In prior experiments at INEEL on samples less than 15 cm (6 inches) in diameter, simulated wastes have been typically hot pressed in argon at 20,000 psi and at 1050 °C (higher temperatures have been used and may be needed depending on the waste composition). The potential consequences of HIP facility accidents are significant because of the pressures and temperatures involved in the process, and these risks from HIP operations were not evaluated in the site risk assessment (Slaughterbeck et al., 1995). The engineering and metallurgical problems that would be encountered in designing a large-scale HIP facility to meet applicable safety standards and codes are probably too large to make HIP a viable option.[3]

2. A second factor is the estimated low-output rate for producing a waste form by HIP. Even with several HIP production lines operating simultaneously, a relatively low-output is anticipated since HIP, as with the HUP process described previously, is a batch process. The possibility of scale-up [50-cm (20-inch) diameter samples have been mentioned] is limited by the problem of finding materials from which large molds can be made (to operate at high temperatures and pressure), as well as the increased safety risk presented by the larger volume of compressed gas as the size of the pressure chamber is increased for higher output.

3. Finally, the HIP process is inherently more expensive than the cold- or hot-pressing processes mentioned above.

---

[3] To expand on the safety distinction between the HUP and HIP processes, HIP requires gas for pressurization because of the high temperatures involved, whereas HUP can be done under high pressure applied hydraulically with metal pistons. This latter method of applying pressure is much safer because of the much lower stored energy of the contents under pressure, as represented in the product of pressure and volume. Although the pressure is comparable in both cases, the volume change after expansion is small for liquids but large for gases. The committee concludes that the HUP process could be safely performed in a hot cell, without the pressurized gas in a HIP process that would pose an additional safety hazard.

TABLE 7.1 Comparison of a HIP Sintered Glass Waste Form Produced at the University of New Mexico to Partially Vitrified (Glass-Ceramic) Waste Forms Produced at INEEL

	Sintered glass produced via HIP at the University of New Mexico, using a nonradioactive surrogate to simulate INEEL calcine	Glass-ceramic produced via HIP at INEEL
Additives	Amorphous silica or/and Na_2O	Amorphous silica and MgO
Waste loading	Up to 50 wt%; demonstrated higher loading is possible	60 to 80 wt%
Processing conditions	Pressure: 1 to 30 Mpa Temperature: 750 to 850 °C Hot isostatic (could also use hot uniaxial) pressing	Pressure: 138 MPa Temperature: 1000 °C Hot isostatic pressing
Processing experience	Prototype plant in Karlsruhe, Germany; up to 30 cm glass cylinders	Laboratory-size samples only
Waste components	1. Most waste components (CaO, Na_2O, Fe_2O_3, MgO, P_2O_5, and B_2O_3) are completely dissolved in the glass phase. 2. Al_2O_3 and ZrO_2 are embedded in the glass phase and occur as crystalline phases. 3. CaF_2 present as undissolved fluorite.	1. Waste components partially occur as calcine relic (baddeleyite, fluorite, etc.). 2. New crystalline phases form, e.g., apatite, zircon, greenockite, Ca-Mg borate, plagioclase, etc. 3. Components incompatible with ceramic phases partition into the glass phase.
Glass phase	1. The glass phase forms a continuous matrix to hold dissolved waste components. 2. Crystalline phases are completely embedded in the glass matrix. 3. Glass phase accounts for >50 wt% of the waste form.	1. The glass phase occurs as discontinuous islands and ceramic phases dominate the waste form. 2. Glass phase is 0-22 wt% of the waste form.
Chemical durability	Forward rate for Na is <1 $gm^{-2}d^{-1}$. Chemical durability determined by the high silica glass matrix (65-75 wt% SiO_2).	Forward rate for Na is up to ~37 $gm^{-2}d^{-1}$ Chemical durability determined by hydrothermally unstable phases, e.g., suanite (6.6-15 wt%) and nepheline
Microstructure of waste form	Simple and easily manipulated.	Complicated and difficult to manipulate and to characterize

NOTE: Prepared by W. Lutze, University of New Mexico, 1998.

Based on the information available, the committee sees no future merit in the HIP process as a practical method for processing the large amounts (volumes) of calcine located at INEEL. The committee knows of no identifiable reason for DOE to continue investigating the HIP process.

Embedding Calcine in Glass

This procedure (Sombret, 1998a; Sombret, 1998b; Bonniaud, Labe, and Sombret, 1968) involves mixing the existing alumina or zirconia calcine with glass particles ("frit") and melting the mixture in a suitable furnace. The partially vitrified end product would consist of particles of the unmelted calcine distributed and embedded in a chemically durable glass matrix. Anticipated advantages of this technique are a high waste loading estimated at 50 weight percent or more, a high throughput similar to that of full vitrification processing, and no requirement for high pressure steps. Unresolved issues are the lack of information for the flow characteristics of the molten mixture, the potential settling of the unmelted (higher density) calcine in the furnace, and the acceptability of a partially vitrified waste form.

Synthetic Rock Waste Forms

A collaboration of investigators at the Lawrence Livermore National Laboratory and the Australian Nuclear Science and Technology Organization has adapted a synthetic rock (Synroc) medium (with a titanate pyrochlore host phase) for potential use in immobilizing excess weapons-grade plutonium. The material produced has a long-term durability (with the appropriate time scale set by the 24,000-year half-life of plutonium-239). In this material, the chemical recovery of plutonium is difficult enough to provide a measure of proliferation resistance.

This Synroc approach in principle could be applied to the INEEL HLW calcine (Godfrey, 1999; Jostsons et al., 1996, 1997). The relatively low (i.e., compared to the characteristics of DOE HLW inventories at Hanford and Savannah River) sodium content and relatively high aluminum and zirconium content of the INEEL HLW calcines are compositional features that would favor good product stability and high waste loading. However, any application of Synroc technology to the INEEL HLW calcines would require development of a process that could accommodate the compositional variations of INEEL calcine. As with all of the nonvitrification options discussed in this chapter, qualification of the waste form would be required.

SINGLE-USE MELTERS

If a partial vitrification waste form (i.e., one consisting of crystalline particles embedded in a glassy matrix) is considered acceptable for disposal, then a "single-use melter" offers some advantages over the continuous melters now in use. In this concept, a mixture of calcine and powdered glass frit would be partially melted in a container (i.e., the interior crucible) that could become part of the final waste container.[4] Some of the advantages offered by this concept are as follows:

---

[4] This "single-use melter" concept differs from the "in-can" melting concept that was designed in the 1960s.

1. Much higher waste loadings could be achieved than are possible for a fully vitrified product made in a continuous melter. The viscosity and conductivity of the melt and the formation and settling of any precipitates are of major concern in a continuous melter, in which the melt must flow properly to exit the furnace. In contrast, waste loadings of 50 weight percent or more could be expected in a single-use melter, since the objective is only to form a viscous mass of unmelted solids surrounded by a glassy matrix that has an acceptable chemical durability.

2. Higher temperatures could be used, if needed, to accelerate the melting process (i.e., increase solubilities to shorten the melting time) since the service life of the refractory would not be a critical factor. Although higher temperatures cause (a) more rapid corrosion of the furnace, resulting in a shorter operating life, and (b) increased volatilization of species such as Cs and Hg, these disadvantages are not critical in the single-use concept.[5]

3. Shorter "processing" times—on the order of a few hours as compared to times greater than 48 hours now used in continuous melters—are possible because melting requirements are reduced. To meet the objective set forth in number 1. above, it is not necessary to melt chemically durable components such as alumina or zirconia, but to merely surround them by glass. Shorter processing times hold the potential for faster throughput depending on the design of the single-use melter.

4. The relative insolubility of alumina and zirconia in glass does not pose the same problem in partial vitrification as it does in full vitrification. The host phase(s) should be able to accommodate a volume fraction of undissolved oxide and still possess adequate durability. Depending upon the detailed microstructure, separate phases would be present, which might affect the physical and chemical durability of the waste form. For example, localized stresses in the event of thermal-expansion mismatch are possible, and their effect on waste form performance would need to be explored by suitable testing.

5. Greater ease in handling an accident scenario. If something goes wrong with the melter, it is easier to remove a broken crucible and refractory container or liner from a batch furnace than to attempt repairs to a continuous melter.

SUMMARY

All of the waste forms described in this chapter are composed of several glass or crystalline phases, as opposed to the single glassy phase for a fully vitrified product. A blend of the INEEL calcines could probably be used successfully in its present state (i.e., with no dissolution and chemical processing) to achieve waste loadings of greater than 50 percent by weight in a glass/ceramic waste form whose chemical durability would satisfy those criteria now in place to qualify HLW forms. Glass/ceramic waste forms offer potential advantages of higher waste loading, ease of processing, and potentially lower costs than options involving chemical processing and complete vitrification. Whether these potential advantages can be realized will depend on regulatory and policy decisions for acceptance of heterogeneous waste forms in permanent repositories. That is, a potential disadvantage of the heterogeneous waste forms discussed in this chapter is the unknown cost and time to qualify them for permanent storage. This disadvantage may be more of a perception than a real technical issue, insofar as (1) the performance of heterogeneous waste forms can in principle and in practice be assessed, and (2) vitrified waste forms such as the HLW glass made at Savannah River are also hetero-

---

[5] For example, Canadian experiments in the 1960s used aluminosilicate ("nepheline syenite") glass for immobilizing HLW (Lutze and Ewing, 1988, and references therein). The in-cell meltings were performed in mullite crucibles at relatively high temperatures (1500-1600 °C) to give products that are far more durable than the present generation of borosilicate glasses.

geneous due to recrystallization upon cooling. The conclusion is that most, if not all, waste forms are heterogeneous, and insofar as the Savannah River glass formulation cannot be used on INEEL HLW due to its significantly different composition, qualification of the INEEL HLW form will have to be done regardless of which waste form is produced.

8

Tank and Bin Closure

The 11 tanks that contain the sodium-bearing waste (SBW) liquids and the seven bin sets that contain the solid high-level waste (HLW) calcine will eventually be closed regardless of the final disposition of these wastes. This chapter evaluates the closure options for these two storage systems. Because liquids in storage are not considered a long-term solution for disposition, the tanks will be "emptied" before closure. Storage of the calcine solids in the bin sets have potentially a longer term stability than tank storage of liquids; therefore, the bin sets may be considered as either "emptied" or full when closed. The quotation marks around emptied emphasize that the bins and tanks will contain some residual waste due to limitations of the present drawdown systems and the difficulty of achieving complete decontamination. These storage systems are presently monitored continuously and actively controlled, and it is assumed that such monitoring and control will continue for several decades while final disposition is being carried out. The starting points for the site's evaluation, as contained in a bin set closure study (Dahlmeir et al., 1998) and a tank closure study (Spaulding et al., 1998), are the following:

- The 11 tanks will be emptied to the level of capability of the existing steam jet and airlift systems leaving a residual 4- to 12-inch layer of highly acidic heel (probably a mixture of both solids and liquids) at the bottom interspersed in the cooling pipes. Deposits on the walls behind and on the cooling pipes will be present also (Spaulding et al., 1998).
- The bin sets will have the calcine removed to a capability of the presently planned retrieval method (see Chapter 2). Transport piping and the existing 8-inch risers used in the retrieval task are the hardware available to remove calcine prior to closure. Additional secondary containment will be in place to enable new openings to the bin sets to be made to carry out further decontamination activities as needed (Dahlmeir et al., 1998).
- The regulations impacting on the technical issues are those currently in place, although changes are likely and new regulations may be promulgated before closure is initiated (Spaulding et al., 1998; Dahlmeir et al., 1998).

The interaction between the technical issues of practicality and regulatory issues of what is sufficient will always be dynamic. Opportunities for trade-offs exist, and the Department of Energy (DOE) and its contractors should identify, define, and evaluate these trade-offs in order to permit an understanding of the possible benefits and risks among several options being considered for closure.

Closure of the two storage systems is evaluated separately in the following discussion; however, there may be common technical issues for both systems. There are three technical actions involved in the closure process:

1. characterization—knowing what remains,
2. decontamination—knowing what added effort at retrieval is needed, and
3. stabilization—how to immobilize the residual waste.

The closure of both of the two storage systems (i.e., tanks and bins) is evaluated for each of these three actions.

TANK SYSTEMS: CHARACTERIZATION

A heel composed of both solution and sludge deposits is anticipated in the bottom of the tanks (Spaulding et al., 1998). The present characterization data on solutions in the tank are available for process control and engineering design purposes, but these data are probably not a reliable predictor of the composition of the bottom heel, because there may be an enrichment of hazardous elements and radioactive isotopes in the sludge portion (Spaulding et al., 1998).[1] To determine the level of decontamination and the degree of immobilization for closure, characterization of the heel is essential. The tank sludge has accumulated through progressive processing campaigns; therefore, old analyses are probably not relevant to current conditions. For this reason, old analyses (>5 years old) of the composition of the SBW sludge were mainly discarded (Garcia, 1997). The remaining analyses are sparse and vary considerably from sample to sample.

The characterization database does not describe how the samples of sludge are obtained. Sample size appears to be a problem issue. Larger sample sizes involve more risk to personnel and make handling more difficult than smaller sample sizes. In this vein, microanalytical methods to assay just the hazardous and radioactive constituents needed for closure could be considered. Variations between samples are expected to be large compared to the instrumental accuracy; therefore, reliance on a sufficient number of small size samples (as opposed to the large sample sizes) may improve the overall characterization effort. Sampling of sludges at other DOE sites is a challenge also, and the Idaho National Engineering and Environmental Laboratory (INEEL) should continue to capture any benefits from these other investigations.

The tank closure study (Spaulding et al., 1998) identifies a task to characterize the heel (presumably including sludge) in all but one of the six options for tank closure (Spaulding et al., 1998). Two of the options characterize each tank over a 1-month period, 6 months prior to initiation of heel removal. Heel removal of all tanks is spread over about a decade starting in year 2007, and removal is done one tank at a time. Three options start characterization of heels in year 2000 and continue until 2016. Some options are premised on a risk-based analysis before qualification and acceptance by regulators. A risk assessment requires characterization to identify the source terms of the hazardous and radioactive constituents. The current characterization database is insufficient to define the risk unless large margins of conservatism of the key constituents can be justified without a heel analysis. Ideally, all eleven

---

[1] Calculations of radionuclide inventories for tank closure use data from Garcia (1997). There is no traceability from the original sources.

tanks should be sampled and characterized before selecting a closure option. If a tank heel can be sampled and characterized in 1 month, then the sampling and characterization schedule for all eleven tanks could extend over a period of 2 to 3 years or more, depending on the ease of deployment of sampling equipment and the analytical laboratory capacity.

Characterization of the tanks has the goal of providing data for the following two goals:

1. process control of liquids that are removed by the existing steam jet and airlift for subsequent treatment (i.e., parameters needed for the design and testing of chemical process steps used in the downstream processes to treat the liquid waste), and
2. risk assessment evaluations to leave or retrieve the remaining heel.

These two goals may allow for different protocols to be developed in sampling and in the use of analytical resources. For example, process control analyses may need large sample sizes of solution to analyze for major constituents that define downstream adjustments of process parameters or chemical additions, and large liquid samples might then be collected for this purpose. However, the risk assessment may only need smaller samples sufficient in size to identify the existence of hazardous and radioactive species in the sludge and their concentration relative to the solution. For the solution portion of the heel, this need may be met by analysis of a sample collected for process control purposes. In this approach, the sludge (i.e., the solid part) of the heel would be the primary focus for near term sampling development.

Important answers provided by sludge sampling would include the following:

- the content of Resource Conservation and Recovery Act (RCRA) materials such as heavy metals (i.e., Hg) and chlorinated hydrocarbons that were used at the INEEL site and that, if present in the tanks, would likely settle to the bottom;
- the total organic carbon content, which could be tested by a procedure such as the carbon dioxide released on ignition of sludge;
- the curie content of alpha emitting isotopes and their spatial distribution in the heel;
- the curie content of beta/gamma emitting isotopes; and
- the mass of sludge per area of tank floor surface.

Sampling and analysis of sludge to obtain this information would probably settle the issue of establishing a "representative sample" to satisfy regulatory and stakeholder needs. Sludge sampling has been done in the past, so it is possible and practical. In collecting sludge data for performance assessments and other uses, the degree of analytical rigor that is required should be assessed to determine whether innovative thinking would permit simpler procedures to be used. Full advantage should be taken of any regulatory flexibility.

TANK SYSTEMS: DECONTAMINATION

Decontamination of a tank involves removing as much as possible of the liquid and sludge portions of the heel on and below the piping array. Six options for decontamination of the heel have been studied. One involves complete ("clean") removal of the heel and dismantlement of the tank and vault. The remaining options involve decontamination to an extent sufficient for a regulated closure based on a risk assessment. The tank walls and cooling pipes are washed down to remove suspected sludge deposits. The remaining heel that is not removable by steam jets is reduced from about 12 inches off the tank floor to about 1 inch using a disposable peristaltic pump. It is not evident how much energy[2] is needed to suspend the sludge for removal.

The preferred approach of Spaulding et al. (1998) appears to be a grout sweep-out system that directs the residual heel to an existing steam jet or airlift system. The process involves using a grout slurry that lifts and displaces the solution, and presumably the sludge as well, so that it raises the liquid to the level of a steam jet. The grout slurry is assumed to capture the sludge. A maneuverable remote transfer arm will be placed inside the tank to deliver the grout to the entire floor area. The design must consider the cooling pipe array and the brackets holding the pipes that may resist flow of the slurry. The sweep-out system may be evaluated in a mock-up facility before implementation (Spaulding et al., 1998). The Savannah River Site has grouted a tank in this fashion but in a different environment and under different regulatory rules (Spaulding et al., 1998).

Success of this process depends on the reliability of mock-up testing. Unsuccessful testing will raise concerns and extend the development costs. If the grout solidifies prematurely in any of the transfer systems, recovery measures will have to be pretested as part of the development. Most tanks have two to three injection ports, except for one tank, which has only one port (Spaulding et al., 1998).

The conditions for grouting the sludge in the tank are uncertain at present. Preliminary testing has shown that the pH of the heel is an important criterion for making a strong grout (Spaulding et al., 1998). To adjust the pH to between 0.5 to 2.0, the heel will be diluted by a series of washdowns. During this dilution, additional precipitation is expected to add to the sludge. There are concerns that sludge deposited under highly acidic (>2 M) conditions and sludge deposited under lower acidic (near 0.5 to 1.0 M) conditions may have different mixing qualities with the grout slurry. In any case, the degree and quality of mixing is not measurable when performed in situ. With no assurance of a well-mixed and strong slurry, the release rate of hazardous and radioactive constituents by the grout will be in question, as will the ability to model this release rate conservatively. Valid and reliable sludge simulant experiments may be required, including large-scale mock-up testing to certify the reliability of the concept.

The criteria for the acceptable degree of decontamination of a tank should be established before implementation of a technical approach. If the residual sludge can be reduced sufficiently, the certification of the grout/sludge mixing may be unnecessary for a given level of risk.

TANK SYSTEMS: STABILIZATION

The degree of stabilization (e.g., via addition of grout) needed for the hazardous chemical elements and radioactive isotopes remaining in the tank is determined by the mobility of these species in the environment after the loss of physical containment (e.g., a breach in the tank wall by corrosion). After closure there is a period in which one or more barriers is the key means for assuring continued containment of the residual waste. During this period some radioactive elements decay, and those that remain to contribute to the time-dependent activity are elements that can be stabilized to varying degrees. For hazardous elements such as mercury or heavy metals, the hazard remains forever; therefore, their long-term

---

[2] An example of one of these energy-adding techniques is to bubble an inert gas into the tanks while the liquid is drawn off, in order to agitate and suspend solids in the liquid to expedite their removal.

concentrations in the surrounding environment, as determined by their rate of release from stabilized residues and their mobility, should be within accepted regulatory standards.

The regulations governing tank closure do not completely specify the technical means for compliance. Although regulatory procedures often do not use risk assessment calculations as a means for establishing a safe condition, risk assessments provide a rationale for trade-offs among various technical considerations. Two such considerations are the degree of decontamination of the tank interior that is to be achieved and the requisite quality of mixing between injected grout and residual sludge. These two issues are related—for instance, if the grout does not mix with the sludge attached to the steel tank surface (the primary environmental barrier), then no credit can be derived from the grout, and the decontamination specification would need to be correspondingly more stringent. Some options for tank closure after stabilization of the grouted heel include continued void filling with grout made from uncontaminated materials or from materials containing one or more substances that are hazardous or radioactive. If the tank void is filled with these substances, the degree of stabilization of the original heel may be irrelevant.

Comments are given below about stabilization of specific constituents that remain in the tank after closure.

Water

A primary requirement for closure is that the tanks contain no free liquid. If the tank is grouted, the grout usually bonds the water as a hydrated species, and the water stays bonded under containment at ambient temperatures. However, the water may become mobile if there is a thermodynamic driving force such as temperature gradients in the tanks, which could, over long periods, allow free water to collect in localized areas within the tank. Thus, thermal gradient testing may be as important as thermal cycling.

Halides

Halides (chloride and fluoride) are potential accelerators of intergranular stress corrosion of stainless steel. Aluminum nitrate is used to complex the halides. Thus, materials used to fill the tank voids must be monitored for high salt content.

Mercury

Until there is an understanding of where the mercury (used in spent fuel reprocessing) is distributed in the waste streams, it will be important to monitor the mercury concentration in grout, and the effect of this mercury contribution on the performance of the system upon closure.

Radioisotopes

The standards for the U.S. Nuclear Regulatory Commission (USNRC) waste Classes A and C define the disposition of allowed radioactive waste in a shallow land disposal facility. The isotopes of concern are ^{90}Sr and ^{137}Cs in the early period when containment is the principal method to control dispersal. After a few hundred years, these isotopes will decay and the overall activity decreases substantially. Thereafter, the principal isotopes contributing to the activity are alpha emitters such as 239Pu and 240Pu and a beta (β) emitter, 99Tc. It is important to know if these isotopes are concentrated in the sludge part of the heel in order to show that stabilization meets the performance goals derived from applicable concentration and release limits. After loss of containment, the 99Tc leach rate may be the key parameter in a risk assessment.

BIN SET SYSTEMS: CHARACTERIZATION

The amount of calcine in the bins after decades of storage has been estimated from calculations of calcine added to the bins from a series of processing campaigns (Dahlmeir et al., 1998). If mechanisms such as caking have occurred to make the calcine stationary, operations such as characterization and retrieval might have to rely on an in-situ comminution process.

In-situ sampling of this calcine is difficult and may not be useful unless the bin contents have been homogenized[3] sufficiently before calcine removal, which seems impractical and unlikely. Access mechanisms into the bin systems are very limited so that obtaining representative samples of the residuals would probably require extensive alteration of the monitoring and control systems. Consequently, characterization of the calcine composition prior to retrieval would be a challenge.

As a result of this challenge, the proposed closure program for the bin sets (Dahlmeir et al., 1998) does not include a significant characterization task for unretrieved calcine. Risk assessments of the options for closure will rely on characterization of retrieved calcine and on conservative bounding estimates of the residual calcine left after retrieval. The proposed strategy of Dahlmeir et al. (1998) is to apply a fall-back sequence to closure using best-technology efforts that start with an attempt to achieve "clean" closure first. If this is not attainable, then Resource Conservation and Recovery Act (RCRA) closure and, finally, closure to "landfill standards" criteria would be applied.

BIN SET SYSTEMS: DECONTAMINATION

Decontamination of the bins depends on the success of the airlift transport system for removal of the calcine. A planning assumption is that 5 percent of the calcine will remain in the bins after completion of the retrieval program (Dahlmeir et al., 1998). Cylindrical bins and annular bins may require different approaches to air lifting of the residue. There are also internal supports in the bins that may trap calcine.

As described in the characterization section above, the bin sets have limited access points, and the monitoring and control system would probably require alteration before any in-situ decontamination operations could occur. However, alterations of hardware to properly satisfy eventual closure would best be done before the start of calcine removal operations, and actions taken during calcine retrieval will impact the success of bin decontamination. Therefore, the retrieval and bin closure operations are related, and the two engineering projects—the Calcine Retrieval and Transport Project (CRTP) and the Bin Set Closure Project (BSCP)—responsible for these operations have a potentially complicated interface.

---

[3] The previous chapter considered "sufficient" homogenization of already retrieved material to ensure product consistency in physical or chemical treatment processes. This chapter also considers waste homogeneity, but in relation to sampling stored waste prior to retrieval. In particular, in situ characterization is of interest for the residuals likely to be unretrieved and therefore to remain in the tanks or bins during their closure.

As an example, CRTP actions can influence the ability to decontaminate the bins (Spaulding et al., 1998). The rush to get CRTP operations under way to retrieve calcine from the bins, to feed downstream HLW and transuranic (TRU) separations, may put pressure on funding and management to ignore the closure task. This situation splits the system development activity and narrows potential innovations to a subsystem perspective. The system function is to move the calcine, which has three subfunctions of retrieve, transport, and decontaminate. Since airlift mechanisms for both CRTP and BSCP are only in the conceptual stage, a closure cycle system outlook is important. Because the lift mechanisms during retrieval and decontamination have the same principle of operation, the proof-of-principle and mock-up testing can be developed in concert, rather than under separate project requirements. A similar synergism is to apply the energy dissipation of the decontamination mechanism (e.g., a CO_2 decon system) to the retrieval operations, where this energy may be useful to solve the problem of calcine inhomogeneity in the bin during the retrieval phase and to break up potential caking.

Clean closure of bins requires nitric acid treatment. Undissolved calcine is expected to remain after treatment so that more aggressive treatments may be necessary, but their effectiveness may be questionable. The calcine has a much higher specific activity than the SBW in the tanks and the bins were not designed for transport of liquids. As a result of these physical conditions and high radiation fields, the intensive decontamination requirements associated with clean closure of the bins are considered by Dahlmeir et al., (1998) to be impractical.

BIN SET SYSTEMS: STABILIZATION

As stated above, the clean closure option for the bins has been deemed impractical (Dahlmeir et al., 1998). Therefore, the approach for defining closure specifications is based on a risk assessment of the residual calcine left in the bin. The assumption in Dahlmeir et al. (1998) is that any hazardous constituents will be delisted (see Chapters 9 and 10 for further discussion); hence, the risk is determined by the stabilization of residual radioisotopes. As with the tanks in the tank farm, there is an incentive to fill the bins with low-level radioactive grout that contains radionuclide concentrations at USNRC Class A or C levels. The option for closure becomes dependent on the acceptance criteria, qualification of source data, and algorithms used in the assessment. This strategy requires a credible characterization of the remaining calcine, which is not now planned. It is possible to provide a conservative bounding of the key characterization data to show that the risk is acceptable in lieu of accurate measurements. Selection of a particular bin closure option appears possible using this risk-based approach.

To qualify for filling with clean grout or grout containing radioisotopes at Class A or C standards, an upper limit on the amount of residual calcine remaining in the bin sets must be known. Further, the residual calcine must be well mixed with the grout to receive credit for its stabilization. These issues are similar to those for tank farm closure.

The curing of the grout in the bin environment will have to be prequalified for leachability, coherence, and strength, since it cannot be tested in situ. Assurance that no free liquid remains after a 7-day cure would also be needed.

The full set of trade-offs between exposure to personnel and risk to the public have yet to be assessed. This situation apparently requires a negotiation with the Idaho state regulators (Dahlmeir et al., 1998).

USE OF A RISK ANALYSIS TO DEVELOP CLOSURE SPECIFICATIONS AND TO MODEL RELEVANT RISKS

A probabilistic risk assessment was done (Slaughterbeck et al., 1995) to evaluate the health and environmental risk from current INEEL operations and future waste treatment options. The risk assessment was a high-level analysis and did not focus on specific designs for future facilities. It evaluated consequences of events such as fires, earthquakes, explosions, and an aircraft accident. Risks to the public and to on-site personnel 100 meters or more from the origin of the accident were evaluated. All the options considered in Slaughterbeck et al. (1995) included tank and bin closure. The risk from tanks and bins after closure was assumed to be nonexistent. Within the limits of the Slaughterbeck et al. (1995) analysis of accident scenarios, all waste treatment options for INEEL HLW, including the "no action" option, have similar risks. This is due in part because of the limitations of that analysis, which did not model in detail the operations internal to any plant facility, did not intercompare various treatment options, and did not consider risk to workers in close proximity to such operations. The committee found that not enough information about the various technical alternatives is available to detail the risks associated with each. In general, however, the more complicated the process steps that are conducted, the greater the technical risk.

A risk analysis of different scope, to model long-term environmental effects and processing risks for workers, could and should be done to develop closure specifications for the tanks and bin sets. One potential outcome of such a site risk analysis is that all treatment options may have similar public risk, but that risk to personnel will be lowest for the closure option with the least manpower activity involved. If the calcine is retrieved and separated into fractions that are stored and/or disposed of at the site but not in the bins or tanks, then one would expect little change in public risk. If the closure assumption is that as much as 5 percent of the calcine may be unrecovered by retrieval and separation for offsite disposition, then one must ask how much increase in public and personnel risk occurs if 100 percent of the calcine is left. In other words, is the current bin storage of calcine acceptable on a relative risk basis, and can treatment options be justified in risk terms?

One way to probe these issues is to ask the following questions in the event that 100 percent (versus 5 percent) of the calcine is left.

- Is the health and safety risk to the public increased 20 times?
- If there is a loss of containment (i.e., a leak), does the dispersal rate increase by a factor 20?
- Is a factor of 20 within the margin of conservatism of data and algorithms used in the assessment?

Many risk analyses work with margins of a few decades; hence, a factor of 20 is a relatively modest one.

If 100 percent of the calcine is left in the bins, rather than 5 percent, monitoring and controlling operations should not result in 20 times greater risk from radiation exposure because the calcine is self-shielding. The calcine has been safely contained for decades and will be monitored and controlled for at least several more decades. If there is a gradual degradation of containment (e.g., a rapid rate of corrosion of the steel vessel, which has not been observed yet), forecasts of life expectancy should be possible on at least a several-decade timeframe. The structural integrity of the bin sets, improved and updated as needed, should protect against an earthquake during the monitoring and controlled period. In the meantime, the radioactive sources are decaying and risk of radiation exposure is decreasing. The conclusion is that whatever the risk level is now to workers and the public, it is decreasing and can be managed adequately for the foreseeable future.

Similar risk-based arguments can also be used to derive tank closure specifications and strategy. Cursory analysis of the content of the calcine indicates that the principal contributors to risk (e.g., heavy metals and actinides in the heels) are the same as for the SBW heels in the tanks. The integrity of the containment vessel, and monitoring for releases, would be important components of the strategy to control risk in the next several decades. After the integrity of containment can no longer be assumed, releases to the environment would be controlled by the waste form (from which radioactive and hazardous chemical constituents would leach into the surrounding environment) left in the tank. This would be in combination with other factors such as environmental mobility and rainfall that are less amenable to direct control.

The bin sets in their present form have pressure relief valves to release gas if it accumulates in the bins. Characterization of samples of calcine show that gas evolution in sealed containers occurs when the calcine is heated above 200 °C due to decomposition of the nitrate (Garcia, 1997). Over time, the temperatures from internal heat sources will diminish so that heating would have to occur from an external source. Closure would require containment of gas as well as solid when the facility goes to a passively monitored state from an actively monitored and controlled state. The risk analysis would need to include a potential pressure event. The degree of severity depends on the free space in the bin and the amount of calcine retained.

9

Constraints Imposed by Disposal Options, Regulations, and Cost

A major consideration in deciding *whether* to process any radioactive waste for long-term conditioning is that of the *risk(s)* being mitigated. The fundamental purpose of environmental regulations [such as those of the Resources Conservation and Recovery Act (RCRA); the Comprehensive Environmental Response, Compensation, and Liability Act (CERCLA); and U.S. Nuclear Regulatory Commission (USNRC) directives] and radioactive waste policy legislation is minimization of risk to human health and the environment. In the absence of other clear and concise criteria and guidance, a driving consideration in deciding on a radioactive waste management strategy should be the comparative risks of the alternatives being considered (including those of limited or no processing). Comparative risk assessment calculations provide quantitative information about risk reduction strategies, and these calculations are required to choose among alternative approaches in a fully informed and objective way.

Once a decision to process waste has been made, key issues[1] to be considered in choosing among treatment alternatives include the following:

1. applicable regulatory requirements, such as RCRA requirements on hazardous chemical constituents, U.S. Environmental Protection Agency (EPA) maximal achievable controlled technology (MACT) rule restrictions on mercury and other effluents, and transportation regulations;

2. disposal options, that is, the availability of present and possible future repositories and their associated waste acceptance criteria;

3. worker and public safety, particularly for process steps with potential hazards such as radiological exposure;

4. technological risk, that is, the probability of technical success of a sequence of process steps, particularly if the limits of existing technology are taxed; and

5. considerations of total life-cycle cost.

---

[1] This is not an exhaustive list of all relevant factors. For example, one nontechnical factor not treated in this report is that of Department of Energy (DOE) budgetary considerations, which usually favor relatively constant spending from year to year and discourage large capital outlays over a short period.

Integral to a decision to select among processing options is consideration of the types and quantities of the final waste form products, whose classification and material properties (e.g., chemical durability) will affect, at least in part, how these five issues are handled.

This chapter addresses issues 1, 2, and 5. Issues 3 and 4 are not discussed in this chapter because they have been treated in the context of specific process steps in the preceding chapters, with further comments offered in the chapters to follow. This chapter begins by providing an overview of the regulatory and other legal requirements (issue 1) for which the Idaho National Engineering and Environmental Laboratory (INEEL) high-level waste (HLW) program plans must account. The chapter continues with an overview of disposal options (issue 2) because of the importance of an available disposition pathway to the formulation of program plans. The chapter concludes with a brief discussion of cost considerations (issue 5) that are relevant to deciding among various processing alternatives.

REGULATORY AND OTHER LEGAL REQUIREMENTS

The INEEL HLW contains not only radioactive constituents, as defined and regulated in the Atomic Energy Act (AEA) and its Amendments, but also hazardous chemical constituents, as defined and regulated in RCRA. While these regulations are briefly discussed below, this discussion does not encompass all relevant regulations, such as the EPA's MACT rule providing air quality permitting restrictions.[2]

Radioactive Waste Classification and Consequences

The AEA regulation and classification of HLW requires that the ultimate disposal of the high-level fraction be in a geologic repository, whose waste acceptance criteria must be met. Transport to such a repository requires that all applicable transportation regulations be met, and that states through which wastes travel are willing to allow these shipments (Wichmann et al., 1996; Wichmann, 1998a,b). The full suite of rules regulating radioactive waste classification and disposal are promulgated by EPA rulemakings, USNRC regulations, and DOE Orders, under the provisions of the AEA and its amendments, the EPA charter, the Low-Level Radioactive Waste Policy Act, and the Nuclear Waste Policy Act.

Hazardous Chemical Constituents and Consequences

RCRA regulation requires that mixed (low-level) wastes that are to be disposed of in the shallow subsurface must be either delisted or adequately characterized and treated, by approved methods, for disposal in facilities that meet the terms and conditions of the Land Disposal Restrictions (LDRs). Closure of tanks and bins would be done under guidance provided by RCRA or the EPA's CERCLA.

---

[2] It is beyond the scope and intent of the committee to provide a complete list of all applicable regulations. The consequences of the MACT rule are not discussed in depth here, in part because this proposed rule had not been finalized at the time the committee gathered its information.

Nonregulatory Legal Constraints

Regulatory constraints on treatment options are augmented by nonregulatory legal agreements DOE makes with other signatory parties, and by court orders, which have in the past established schedules and deadlines for the completion of certain tasks. Specifically, compliance agreements have been established with the EPA and the state of Idaho (see Appendix G).

DISPOSAL OPTIONS FOR WASTE PRODUCTS

Any waste forms produced from the treatment and immobilization of INEEL HLW will likely be disposed of, probably after a period of interim storage. The current suite of DOE disposal sites are briefly listed here, with more detail provided in Russell et al. (1998) and Russell and Taylor (1998). Each disposal site, when operational, will have "waste acceptance criteria" (WAC) that specify the requirements that waste forms must meet to be accepted for disposal.

HLW: First and Second Repository Programs

The high-activity waste (HAW) from treatment of INEEL HLW calcine could be immobilized in a waste form suitable for acceptance and disposal at a geologic repository. As specified in the Nuclear Waste Policy Amendments Act of 1987 (P.L. 100-203), the candidate repository at Yucca Mountain is the only site for DOE to currently study and characterize to ascertain its suitability. The Nuclear Waste Policy Act of 1982 (P.L. 97-425) limits this repository to 70,000 MTHM[3] of spent nuclear fuel (SNF) or an equivalent amount of HLW (Office of Civilian Radioactive Waste Management, 1998: p. 13). Because 70,000 MTHM is less than the total projected inventory of commercial SNF (ORNL, 1996), a second repository program (i.e., an extension of Yucca Mountain's legal limit, or another repository) will be needed to dispose of the remainder of the commercial SNF.

To dispose of DOE SNF and other forms of DOE HLW, a "co-disposal" strategy has been proposed (DOE, 1985, 1987) and pursued to date, in which 10 percent of the Yucca Mountain repository's capacity (i.e., 7,000 MTHM) would be reserved for inventories of DOE HLW. Even in this case, because the total DOE inventory of HLW exceeds 7,000 MTHM, a second repository program also would be needed for the remainder of the DOE HLW. DOE plans show the first repository to be filled by 2035, with an undetermined amount of INEEL HLW included in this first repository's inventory (Office of Civilian Radioactive Waste Management, 1998: pp. 7, 10-11; Wichmann et al., 1996: p. 2). Therefore, some, or perhaps all, of the INEEL HLW calcine is one of the candidate DOE HLW streams for disposal in a second repository, particularly if processing is not completed until close to 2035 or later.

---

[3] To be precise, section 114(d) of the Nuclear Waste Policy Act of 1982 restricts the capacity as follows: "The [U.S. Nuclear Regulatory] Commission decision approving the first such [license] application shall prohibit the emplacement in the first repository of a quantity of spent fuel containing in excess of 70,000 metric tons of heavy metal (MTHM) or a quantity of solidified high-level radioactive waste resulting from the reprocessing of such a quantity of spent fuel until such time as a second repository is in operation."

HLW: Waste Acceptance Criteria

The plans for disposal of commercial SNF in the Yucca Mountain repository stipulate use of a waste package container whose design would be tailored to provide adequate long-term containment. Unconditioned spent fuel would be the waste form within the waste package. The DOE SNF and other forms of HLW from defense-related activities within the former DOE weapons complex that would be co-disposed with commercial SNF would be placed in a similar waste package container. Current disposal concepts consider DOE and commercial SNF as unconditioned, with other DOE HLW (such as INEEL calcine) processed into suitable waste forms.

A vitrified HLW form is being produced for DOE HLW at both the West Valley Demonstration Facility in New York and the Defense Waste Processing Facility at the Savannah River Site in South Carolina. The Yucca Mountain WAC (Russell et al., 1998; Office of Civilian Radioactive Waste Management, 1998; DOE, 1996c) are still under development and will ultimately specify the requirements for these HLW forms. Although the HLW form is nominally to be borosilicate glass (DOE, 1996c: p. 7; Office of Civilian Radioactive Waste Management, 1998: p. 21), the current Yucca Mountain WAC make provision for other waste forms to be considered for disposal (Office of Civilian Radioactive Waste Management, 1998: pp. 15-16). To the committee's knowledge, the Nuclear Waste Policy Act (P.L. 97-425) specifies only that the DOE HLW be solidified; therefore, the solid waste form(s) of DOE HLW would seem to be a matter of DOE policy only and if so, use of nonborosilicate glass waste forms would not require any change of law.

Neither a second repository program nor definitive WAC for this repository exists. In the absence of any WAC for a second repository program, one approach could be to use the Yucca Mountain WAC as a surrogate. However, two potential problems arise with the use of the Yucca Mountain WAC to guide the INEEL HLW program.

The first potential problem is that using the Yucca Mountain WAC could be challenged insofar as the current Yucca Mountain WAC (1) may not be proper for a different geologic site, and/or (2) are likely to change over time.[4] In support of the potential for change, the current "Systems Requirements Document" (Office of Civilian Radioactive Waste Management, 1998) is on its fourth revision, and the current "Waste Acceptance Product Specifications" (DOE, 1996c) is on its second revision, and the Yucca Mountain site has yet to be licensed by the USNRC. Unduly restrictive or lenient WAC could have significant consequences in setting the processing specifications for the waste form to be produced from the INEEL HLW calcine.

The second potential problem is that the Yucca Mountain WAC exclude RCRA constituents (Office of Civilian Radioactive Waste Management, 1998: pp. 11,19; Wichmann et al., 1996). However, INEEL HLW calcine and SBW have both characteristic and listed RCRA hazardous constituents (Wichmann et al., 1996; Wichmann, 1998).[5] Any disposal of these wastes must involve either of the following:

---

[4] It is possible that another geologic repository site might place different requirements on the waste form and waste package than the performance requirements presently needed for the Yucca Mountain site.

[5] RCRA-characteristic materials are hazardous because of a characteristic (e.g., ignitability, corrosivity, reactivity, or toxicity) that qualifies them as hazardous; examples of RCRA-characteristic materials are "D-listed" materials such as the metals mercury, cadmium, and lead. RCRA-listed species, such as benzene, carbon tetrachloride, trichloroethylene, and other solvents, are items specifically named in the F-, K-, P-, and U-lists. Both characteristically hazardous and listed materials are regulated under RCRA. The difference is that, following suitable treatment, the hazardous characteristic is presumed to be removed, implying that characteristic materials are "out of RCRA." In contrast, listed materials are never "out of RCRA," even though treatment is still required to dispose of them. Their disposal must be in RCRA-regulated facilities.

1. treating them to the terms of the RCRA LDRs, using a method equal to or better than the appropriate "Best Demonstrated Available Technology" (BDAT), or
2. obtaining suitable treatment variances (40 CFR 268; Wichmann et al., 1996) from the state of Idaho.

Following this treatment, the RCRA-characteristic wastes are "out of RCRA" and therefore qualify for disposal in a facility outside of RCRA controls. However, the RCRA-listed constituents must be disposed of in a facility with RCRA controls. The Yucca Mountain repository WAC does not allow such RCRA-listed constituents.

Two solutions to this problem are possible. One is to assume that the appropriate repository for INEEL HLW will comply with RCRA disposal requirements. In the absence of a repository complying with RCRA controls for disposal, the second solution would involve "delisting" the RCRA-listed components (Wichmann et al., 1996; Russell et al., 1998). For this second strategy to succeed, the DOE would present delisting petitions to each state through which the waste would be transported. All these states would have to approve such petitions (i.e., approve of the transport and disposal of the waste in the form in which it is proposed to be packaged) (Wichmann et al., 1996).

Uncertainties Associated with Geologic Repositories and Co-Disposal

In the early 1960s, a waste management decision was made at INEEL to solidify acidic waste by calcination and to store this calcine in bins. At that time the current suite of regulations (i.e., from regulations promulgated since the 1960s, such as from the EPA and USNRC) that apply (or potentially apply) to storage of this kind did not exist. It is of course conceivable, if not likely, that the future might involve further changes in regulatory requirements.

With respect to the capacity of the first geologic repository as well as the development of a second repository, there are many current uncertainties, including how to express the disposal capacity for DOE HLW. This is an issue of equivalency between the measure (MTHM) used for commercial SNF and that for DOE defense waste co-disposed with it. No legal guidance defines this equivalent measure (Knecht et al., 1999); therefore, any change to it would seem to be a matter of DOE policy only and if so, would not require any change of law. The current method (DOE, 1985, 1987) planned for the first repository provides for this conversion by making 0.5 MTHM of SNF equivalent to one Savannah River-size (approximately 0.7 m^3) (Russell et al., 1998: pp. 5, 23-24) canister of glass made from DOE HLW, a mass-to-volume conversion. This implies that DOE HLW waste destined for repository storage is to be measured by volume.

Consequences of Using Various Conversion Methods. As noted previously, DOE plans (Office of Civilian Radioactive Waste Management, 1998, p. 11) call for 7,000 MTHM, representing 10 percent of Yucca Mountain's capacity, to be filled with "co-disposed" DOE HLW rather than by commercial SNF. This 7,000 MTHM has been subdivided into two other "planning" figures: 4,667 MTHM for DOE HLW that is not SNF[6] and 2,333 MTHM for DOE SNF. Using the mass-to-volume conversion introduced above, the 4,667 MTHM of non-SNF DOE HLW is equivalent to 9,334 Savannah-River-size canisters, or approximately 7,000 m^3. This represents disposal in the first repository of less

---

[6] In widely accepted usage, the term "HLW" includes SNF. However, common usage in places such as INEEL that have significant quantities of both SNF and other forms of HLW is to reserve the term HLW only for non-SNF forms. Therefore, most INEEL documents referenced in this report use HLW to refer to the calcine only.

than half (approximately 46 percent) of the projected estimate of DOE HLW that is not SNF (Knecht et al., 1999); therefore, a second repository (or expansion of the first) would be required.

However, other conversions are possible that would affect the amount of DOE HLW that can be co-disposed with commercial SNF (Knecht et al., 1999). One would be to consider that DOE has processed at all of its sites over time approximately 177,000 metric tons of SNF (Kimmel, 1999a), and therefore to count the sum of all DOE HLW as equivalent to 177,000 MTHM. This conversion method would only permit 3 percent (4,667 MTHM of a total 177,000 MTHM) of all DOE non-SNF HLW to be disposed in the first repository, regardless of its volume.

Another possible conversion could be based on radioactivity as measured in curies, using the fact that 1 MTHM of SNF with a burnup of 30,000 megawatt-days contains approximately 300,000 curies (Ci) after 10 years of cooling. As a result, 0.5 MTHM would correspond to approximately 150,000 Ci of HLW (rather than to one Savannah River-size canister). Still another conversion could be based on radiotoxicity, using regulatory release limits in the USNRC's 10 CFR part 20 to compare the long-term performance of commercial SNF to DOE HLW based on how the long-lived radionuclides in each contribute to the radiotoxicity after 1 to 10 millennia (Knecht et al., 1999). In each of the last two conversions, all of the DOE non-SNF HLW is represented by an equivalent MTHM number that is less than 4,667 MTHM. Therefore, in these last two conversion methods, the complete inventory of DOE HLW that is not SNF could be disposed of in the first repository, regardless of its volume (Knecht et al., 1999).

Conclusions. With respect to the capacity of the first geologic repository as well as the development of a second repository, there are many current uncertainties, including how to express the disposal capacity for DOE HLW. Some of these uncertainties—such as the solid waste form for DOE HLW and the equivalency between the measure (MTHM) used for commercial SNF and that for DOE defense waste co-disposed with it—seem to be matters of DOE policy only and if so, adjustments would not require any change of law.

Uncertainty in the conversion of MTHM to an equivalent measure for DOE HLW would affect both the amount of DOE HLW disposed of in a first repository and the importance of its volume. If the DOE defense waste destined for the repository is accounted for not by volume but by some other unit (e.g., the MTHM of reprocessed SNF that it represents, or its curie or radiotoxic content), then its volume will not be the unit of measure for determining how much DOE HLW enters the repository.

The general conclusion is that the current regulatory framework, particularly for the second repository program, is uncertain at present and subject to change in the long term. In particular, this uncertainty tempers the advantages of reducing the volume of the HLW fraction. Although minimizing the volume of defense HLW has benefits in reducing off-site transportation and disposal costs that are dependent on volume, the adaptation of a different measure by which INEEL HLW is to be disposed may make volume reduction a less important planning criterion.

Greater Confinement Disposal Facility

The Greater Confinement Disposal Facility (GCDF), a DOE facility at the Nevada Test Site (NTS), is a disposal site that consists of boreholes into arid alluvium (Cochran et al.,

1999). High-activity low-level waste (LLW) and classified transuranic (TRU) waste[7] have been disposed in these boreholes in the 1980s (Bonano et al., 1991; Cochran et al., 1999). This form of disposal is proposed for consideration for the mixed HLW form generated from any cementitious processing of HLW calcine and SBW (Russell et al., 1998). This disposal concept may have merit, but as noted in Chapter 4, such a use of the GCDF is outside the current regulatory approach for HLW disposal.

Transuranic Waste: Waste Isolation Pilot Plant

The Waste Isolation Pilot Plant (WIPP) near Carlsbad, New Mexico, is the DOE repository for defense-related TRU wastes, and would be a suitable disposal site for INEEL waste streams that meet this classification. The WIPP-WAC specify that both contact-handled (CH) and remote-handled (RH) TRU wastes are to be accepted. WIPP has legal limits on the total volume and total curies that can be accepted in each of these classification categories. The distinction between RH-TRU and CH-TRU, which are terms derived from operational considerations in the handling of TRU waste, is that waste is classified as RH-TRU if the contact dose rate exceeds 200 millirem per hour at the surface of the waste container. For waste to be defined as TRU, the concentration of transuranic isotopes must exceed 100 nanocuries per gram (nCi/g). The WIPP-WAC contain further specifications, such as limits on the total curie content of a waste container, and prohibition of excessive hydrogen-gas generation (via alpha-induced radiolysis of organic material co-disposed with TRU waste) during transport to the site in a sealed package (DOE, 1996a). WIPP can receive mixed (i.e., RCRA) waste according to the terms of the RCRA Part B permit, which was issued in October 1999. The WIPP facility began CH-TRU operations in March 1999.

Challenges Associated with WIPP Disposal

WIPP's suitability as a TRU repository for these wastes would involve resolution of several issues, including the following:

1. Because WIPP accepts only TRU waste, any waste sent to WIPP would have to obtain regulatory approval as TRU waste.
2. Because WIPP can legally accept waste only of defense origin, these wastes would have to be judged as defense wastes to gain entry into WIPP.
3. WIPP is already overcommitted in RH-TRU wastes from all DOE sites; therefore, a legal change is required in order to allow more RH-TRU inventory to be disposed of in WIPP.
4. Adequate consideration should be given to the repository's ability to safely store all the waste components, particularly the gamma (γ) emitters and fission products. For example, ^{241}Am (with a 60 keV γ-ray), ^{233}U, and ^{232}U are three γ-emitting radionuclides that could be contained within some plutonium waste. A suitable TRU repository would provide adequate shielding for workers during waste handling operations and adequate long-term containment given the potential for gaseous daughter progeny. Similarly, the constituents of INEEL TRU waste streams should be examined for their impact on the modeled long-term

---

[7] These wastes are two examples of "special case" or "orphan" wastes because no law governs their disposal. The DOE LLW with activity that is "Greater Than Class C" has no legally specified disposal pathway. The classified nature of the transuranic wastes at the NTS prevents them from legally qualifying for disposal at WIPP.

performance of the WIPP repository. In support of this statement, the EPA's Certificate of Compliance for WIPP in 1998 was based on an assessment of the projected inventory of radionuclides. This issue is mentioned for completeness and is not meant to be alarmist; indeed, the committee believes that there is no problem meeting the technical requirements of the current WIPP-WAC with the TRU waste streams under consideration here.

Non-TRU LLW: Off-Site Disposal

Currently, a few options exist for off-site disposal of a non-TRU LLW stream generated from INEEL. Commercial sites are either precluded from accepting DOE LLW or receive only low-activity waste (Russell et al., 1998). The only currently available off-site DOE options are the disposal facilities at the NTS and the Hanford Reservation.

LLW Disposal at the NTS

The DOE LLW disposal site at the NTS can in general accept LLW from other DOE sites, but currently waste equivalent to the commercial category of Greater-than-Class C is accepted only on a case-by-case basis (DOE, 1997: p. 3-7; Russell et al., 1998, Vol. 1: p. 38). Mixed (i.e., RCRA-regulated) LLW waste is also accepted (DOE, 1997), but the WAC for these wastes are not yet finalized, and wastes generated at INEEL are not approved for disposal there at this time (Russell et al., 1998, Vol. 1: p. 38; Vol. 2: p. 53).

LLW Disposal at the Hanford Reservation

The DOE LLW disposal site at the Hanford Reservation accepts waste with specific activity at or below the USNRC Class C designation (Russell et al., 1998, Vol. 2: p. 56). Certain mixed wastes are accepted (Fluor Daniel Hanford, Inc., 1998[8]). Non-Hanford waste generators can use the Hanford disposal site subject to the approval of the DOE Richland Office (Fluor Daniel Hanford, Inc., 1998).

Non-TRU LLW: On-Site Disposal

On-site disposal of LLW can occur in suitable disposal sites. Currently DOE is self-regulating for the radionuclide constituents, based on DOE LLW disposal criteria that are both general and specific to the INEEL site (Russell et al., 1998; Bonano et al., 1991: p. 56), but the USNRC is increasing its regulatory authority at DOE sites and may be the future regulator. The USNRC disposal regulations (i.e., 10 CFR 61) require that a performance assessment calculation be done to model the fate and transport of radionuclides in the subsurface environment, and to achieve a certain level of performance in the design of the disposal facility.[9] When disposal operations are completed, the facility would be capped and monitored and institutional controls emplaced for 100 years.

---

[8] This is the 1998 version of the Hanford Site Solid Waste Acceptance Criteria.
[9] Specifically, the modeled concentrations of radionuclides released to the environment should be as low as is reasonably achievable, and should not result in an individual whole body dose exceeding 25 mrem/year (10 CFR 61.41).

RCRA-Approved Disposal

Some constituents of HLW calcine and SBW are governed by the RCRA regulation of the EPA. The RCRA LDRs specify disposal site requirements, which include the use of liners and caps for "Subtitle C" landfills. These LDR requirements can be avoided if the DOE were successful in obtaining "delisting petitions" to exempt the waste from RCRA regulatory protocols.

LAW: Non-Geologic Disposal

There is some published guidance for how HLW of sufficiently low activity can be disposed of in a site that is not a geologic repository. A regulatory decision is needed to exempt waste, such as the low-activity waste (LAW) fraction from processing, from further management as HLW.

This problem was encountered at DOE's Hanford site, in the context of which the USNRC provided guidance regarding incidental waste classification for the contents of waste in underground tanks. In the *Federal Register* of March 4, 1993, it was written that, "... the USNRC indicated that spending vast sums of money without expectation of benefit to health and the environment would not be prudent ..." (Wichmann et al., 1996: p.10).

Near-surface disposal is possible if the following conditions/criteria are met:

- the waste activity is less than or equal to the USNRC Class C LLW limit(s);
- a performance assessment shows that disposal will not represent a hazard to public health or safety;[10]
- the majority of the radioactivity will be in a HAW form sent to a geologic repository; and
- technical and economic feasibility is supported, as by a cost/benefit analysis of separations options (Wichmann et al., 1996: p. 10).

Summary and Conclusions on Disposal Options

It is not certain at present that INEEL waste products can be disposed of in most of the disposal options outlined above. This may be because of uncertainties in whether these repositories will be available for INEEL waste products and/or whether the INEEL wastes are properly qualified to meet the WAC. Only the LLW disposal sites, and WIPP for TRU LLW, are operational at present. The only off-site disposal option immediately available for INEEL wastes (i.e., without the need for a regulatory petition or special ruling) is DOE's Hanford site for nonmixed LLW. The committee's conclusion is that, with disposition pathways uncertain, efforts to develop a viable repository option are warranted. The requirements (e.g., repository waste acceptance criteria) so derived from a viable disposal pathway are an important input that should precede any decision among treatment alternatives for the HLW calcine and SBW.

Although HLW from some other sites likely will go to Yucca Mountain (presuming it is eventually licensed and opened), under current restrictions this repository will not accept mixed waste (i.e., containing untreated RCRA constituents) such as the current INEEL HLW calcine. Moreover, Yucca Mountain, if operational, is by DOE plans to be filled by 2035.

---

[10] These performance assessments are part of the requirements for both near-surface disposal of LLW (in the USNRC 10 CFR 61 regulation) and geologic disposal of HLW (in the USNRC 10 CFR 60 and 10 CFR 63 regulations and in the EPA proposed 40 CFR 197 regulation).

Because of these uncertainties in the WAC and in the availability of a first repository, it is reasonable to project that INEEL HLW will be disposed of in a second HLW repository, which is presently totally undefined.

For all non-HLW streams, the USNRC would have to rule on any waste reclassification as incidental waste. TRU waste could go to WIPP, which has recently opened, but which faces capacity limitations due to the amount of defense wastes at other sites. LLW in principle could go to commercial disposal sites in Washington or Nevada, but both of these sites have capacity and regulatory problems to contend with. Whatever the nature of radioactive wastes to be shipped offsite, transportation issues loom as significant problems to be resolved. The INEEL HLW program thus is faced with high uncertainty as to where, and whether, its wastes can be disposed off site.

COST CONSIDERATIONS

At its meeting in Idaho Falls in August 1998, the committee was presented with cost estimates for 27 waste disposal options. These options involved various types and degrees of separation prior to different approaches to immobilization. The present value costs ranged from about $2.5 billion to between $9 billion and $10 billion. The committee gave little credibility to these estimates, largely because of the limited data on which they were based. Many of the cost differences were the result of differing estimates of disposal costs, which are highly uncertain at best. One contribution to the uncertainty of disposal costs is that the disposal locations are not established and developed sufficiently to permit an accurate estimate. Uncertainties as to when actions would actually be taken in the waste disposal options were another unknown that would affect cost figures calculated by discounting to a present value (see Appendix B).

With significant uncertainties represented in the information on which these cost estimates were based, the committee was forced to rely on simplistic, qualitative principles, such as:

- Do not spend money until the task is well defined and the route to its accomplishment identified.
- Low-temperature processes (e.g., evaporation, cementation) are usually less expensive than high-temperature processes (e.g., vitrification and production of ceramics).
- Each additional process step adds to the uncertainty in cost.

While no preliminary numbers were derived for the recommendations in this report, principles of this kind certainly influenced the committee's views.

10

Inconsistencies Associated with Current Plans

The processing alternatives under consideration for the Idaho National Engineering and Environmental Laboratory (INEEL) high-level waste (HLW) calcine and sodium-bearing waste (SBW) will result in final waste forms whose classifications [e.g., high-level, transuranic (TRU), low-level, and/or mixed] and quantities will affect disposal options. Different disposal sites (not all of which currently exist), each with different and largely unknown acceptance criteria, will be needed for the various waste types. The regulatory and legal requirements discussed in Chapter 9 provide other numerous criteria to be accommodated. Designing a waste management and disposal plan to deal appropriately with these constraints is a challenge. In parts of the current plans, inconsistencies in logic arise, as discussed in this chapter.

PROVING THE NEED TO DO ANYTHING WITH THE CALCINE

In the committee's judgment, no waste treatment option should be selected until a substantive risk assessment is done of a no-action or minimal action alternative, particularly since the present calcine storage bins have a design life of approximately 500 years (Palmer, 1998; Dirk, 1994[1]; Schindler, 1974),[2] after which time the radioactivity will have decreased greatly (see Chapter 11). Only one of the bin sets is believed to be "structurally unsound according to seismic criteria,"[3] (Dahlmeir et al., 1998). This bin set does conform to ASME Boiler and Pressure Vessel Codes (Ibid, Engineering Design File "EDF-BSC-007," p. 2), and appears to be equivalent in performance to the rest of the bin sets except that the bins are flat on the ends rather than pan-shaped (Ibid, p. 4).

Basing Decisions on Risk Reduction

Insofar as storage of calcine in bin sets appears to be a low-risk configuration, there is no technical or risk basis, in the committee's judgment, to allocate billions of dollars for

---

[1] This reference estimates corrosion of the interior steel bin walls of only 5 thousandths of an inch ("mils") in 500 years.
[2] 500 years is also used as a design value in Dahlmeir et al. (1998), Engineering Design File EDF-BSC-007.
[3] This quotation, found in Engineering Data File EDF-BSC-002 of Dahlmeir et al. (1998), is from a comment by a bin expert regarding Bin Set #1. This statement is also attributed to Reference 38 of Dahlmeir et al. (1998), cited in Volume 1 on page 78.

processing such low-risk calcine, even if the processing program were low in technical risk. The committee believes that the Department of Energy (DOE) manages much higher risk radioactive waste configurations elsewhere. Therefore, it seems premature to do anything in the near future with the calcine, beyond possibly beefing up the confinement provided by the bin sets (e.g., by improving the protection against seismic events, water intrusion, atmospheric moisture, and wind).

TREATMENT AND DISPOSITION OF SODIUM-BEARING WASTE

The disposition of a low-level TRU waste (namely, SBW) by addition of nonradioactive chemicals (i.e., aluminum nitrate), calcination, and storage with HLW in the bins, thereby converting the SBW calcine to a HLW (via mixing) to create more HLW volume than existed to begin with, is unwise in the committee's judgment, because HLW will likely be more expensive and difficult to dispose of than low-level TRU waste. This approach would prevent the site from exploring TRU waste disposal options for the SBW.

DEVELOPING SOUND PLANS FOR CALCINE TREATMENT

Basing program plans on only two types of calcined materials (Al-type and Zr-type) could misrepresent the full suite of challenges associated with calcine dissolution and treatment, particularly because the relative CaF_2 content is arguably the component that most differentiates the various calcines. The CaF_2 content is potentially problematic because CaF_2 serves as (1) an undissolved solid in nitric acid dissolution and subsequent separations stages and (2) an undesirable ingredient in a borosilicate glass formulation. Another challenge for program plans is the choice to use ion exchange agents only available from either Czech universities or Russian institutes (see Chapter 3).

Reducing Uncertainties, Unknowns, and Assumptions

In the committee's view, a sound strategy would include efforts to reduce major uncertainties, unknowns, and assumptions, as by collecting characterization data and performing tests to validate key assumptions. It would be imprudent to adopt a high-risk, $5 billion-plus operation without collecting calcine characterization data or performing adequate tests. Likewise, at this stage in the process, acquiring data to replace or validate key assumptions is a more informative and useful activity than conducting design studies and cost estimates based on assumptions yet to be validated.

Establishing the Rationale for Separations

One decision in any remediation flowsheet (see Figure 1.5) is whether or not some separations will be done. As stated in the introduction to Chapter 3, a separations approach has merit if the cost savings associated with the reduced volume of final waste can be justified compared to the cost (and risk) of performing the separations. Therefore, reliable estimates are needed on both of these cost figures (and on the risk associated with operations). Unfortunately, one of these cost values depends on HLW disposal costs in the future (i.e., circa 2035), and potentially in a second repository, whose cost structure may differ from the first. Until reliable cost information is available, as also noted in the previous chapter, one cannot assemble reliable figures to judge whether the separations process is a worthwhile approach.

Lack of a Performance Assessment on What Separations are Justified

Performance assessments are useful guides for determining waste form requirements, but such assessments have not been completed (Olson, 1998b). A performance assessment should be done to cover the contingency that the waste cannot be shipped and disposed of elsewhere and to define the requirements for keeping it on-site, particularly because a period of at least interim storage at INEEL is likely. Similarly, a performance assessment should be used to develop requirements on waste form properties (e.g., radioactivity concentrations and leach resistance) needed for shipping and disposal of the wastes at any site. This approach would be more rational than one that just adopted guidelines such as Class A or C limits without regard for the time period during which the waste will be contained. For example, if at least a few hundred years of containment is planned for, whether on-site or in off-site disposal, then ^{137}Cs and ^{90}Sr separations are unnecessary, and the only separations that may be justified are TRU separations.

SELECTION OF SOLID WASTE FORMS

Until the full suite of regulatory requirements for the complete pathway to disposal is defined, important relative advantages and trade-offs between various durable waste forms (e.g., glass, grout, and glass/ceramic) cannot be ascertained. Therefore, the committee believes that it is premature to decide which of these various durable waste forms to make.

In addition, the committee believes it unwise to convert calcine and SBW to forms that would be difficult to dissolve and process further in the future, should that be necessary. For SBW, the alternative is to consolidate it using a less "irreversible" solidification option, such as one of the candidate methods discussed further in Chapter 12. For the committee's views on near-term actions for the calcine, see Chapter 11.

The Need to Develop a Good Waste Form for Defense HLW Co-Disposed with Commercial Spent Fuel

Commercial spent fuel accounts for 90 percent of the contents of the proposed Yucca Mountain repository, as measured in metric tons of heavy metal (MTHM), and 95 percent of the radioactivity as measured in curies. Currently, DOE plans to directly dispose of this commercial spent fuel without processing to produce a more leach-resistant waste form. The co-disposed DOE HLW (representing 10 percent of the 70,000 MTHM legal limit and 5 percent of the projected curies deposited in Yucca Mountain) consists of DOE spent nuclear fuel (SNF) and other HLW forms, such as vitrified HLW from DOE sites. None of these non-SNF HLW forms logically can be expected to have a major impact on the analysis of the risk to the public because of their much smaller contribution to the total inventory of radioactivity. Moreover, some of the longer-lived radionuclides in DOE HLW may have been removed during reprocessing and recovery operations. In the committee's judgment, it is therefore difficult to rationalize requiring durable glass waste forms for these DOE HLW inventories. In this context, it is important to determine the comparative performance of different waste forms for defense HLW that is not DOE SNF.

Therefore, the committee believes it is premature to decide among various processing options (that produce different waste forms) without (1) a statement of waste form performance objectives and (2) a clear and consistent comparison of waste form performance. Although calcine is not currently an accepted waste form for ultimate disposal, there seems to have been no careful consideration of the performance requirements needed and of processing alternatives that achieve these requirements with minimal or no change to the calcine material form. For example, performance requirements might show the present (or slightly modified) form of the calcine to be suitable for direct disposal in a geologic repository (Rechard, 1995). Until this has been proven to be false, the committee believes that there is no rational basis for dissolving the calcine to process it to a different waste form.

Number of HLW Canisters

Reducing the HLW volume and the number of canisters is an important perceived goal, even though the waste may never be shipped off-site. Current cost analyses, using certain assumptions, favor a reduction in volume (Palmer, 1998). By the time waste is shipped in the future, however, the conditions and assumptions that underlie cost calculations may differ. By then, it is not certain whether the transportation and disposal costs associated with the number of canisters would be the controlling cost element.

The right framework for considering disposal volume in a second HLW repository has yet to be established. Therefore, the committee believes that it is premature to base technical choices on waste volume reduction (which may not be important in the future) when other important considerations (e.g., impacts on risk reduction and worker safety, and the degree of technical risk of the proposed plans) are either poorly known or at odds with a strategy to process in order to reduce HLW volume. In particular, dissolution and separation operations on the existing calcine will result in potentially significant risks, costs, exposures, and time requirements. If significant costs and risks are incurred in these operations, the rationale to minimize HLW volume should be tempered with these considerations in guiding the selection of a processing option.

ENSURING A CLEAR DISPOSAL PATH (OUT-OF-STATE TRANSPORT AND ACCEPTANCE) OR ELSE PLANNING FOR CONTINGENCY

A legal and technically viable path should be established for the out-of-state transportation and disposal of the waste products that are to be made. In the committee's judgment, it is premature to develop a waste form without a developed and approved disposition pathway.

In addition, the committee was given the view (Wichmann, 1998) that there is a low probability that regulatory authority would ever be given to move the HLW in any form out of the state of Idaho. The committee has no basis for agreeing or disagreeing with this view, but believes it is prudent for the program to develop a course of action to follow if this proves to be true. While the settlement agreement (see Appendix G) between the DOE and the state of Idaho requires that the HLW be prepared for out-of-state shipment, the program should address the alternative (i.e., retention of the waste in Idaho for the foreseeable future) in its contingency planning.

Resource Conservation and Recovery Act Constituents

As discussed in Chapter 9, the current HLW repository acceptance criteria do not permit disposal in Yucca Mountain of hazardous materials regulated by the Resource Conservation and Recovery Act (RCRA). It seems incongruous to the committee to allow disposal of RCRA-hazardous waste in a contained facility on the INEEL site but not in a geologic repository.

An assumption common to all the INEEL HLW treatment options presented to the committee was that all such plans depend on delisting RCRA waste constituents. These delisting petitions to the appropriate regulatory authorities should be resolved expediently, or else contingency plans should be developed. The committee believes that it would be premature to select a treatment option without resolution of this issue, because the regulatory solution to the RCRA constituents is an important determinant on waste form requirements. Here calcine characterization data would be useful to confirm the concentrations of RCRA materials and the reasonableness of the position that materials such as organics did not survive the calcining process.

The RCRA constituents of the waste impose a potentially significant regulatory challenge. One challenge is in the success of confirmatory sampling and analysis to support delisting petitions for some (e.g., organic) RCRA constituents. For constituents not handled by delisting, some requirements on the final waste form(s) are likely, insofar as the RCRA Land Disposal Restrictions (LDRs) specify disposal criteria. If considerable heavy metals are present, the need to meet Toxic Characteristic Leaching Procedure (TCLP) criteria may impose stabilization requirements prior to disposal.

WHAT SHOULD BE DONE TO RESOLVE INCONSISTENCIES

The committee has identified in this chapter some inconsistent elements in the INEEL HLW program planning, with the intention of elucidating where changes in approach might best be sought. These conflicting factors are presented in this chapter as issues relevant to any future decision about processing HLW calcine and SBW. The committee expects that resolution of these issues will assist any decision on alternative treatment technologies. The next two chapters deal with the committee's views of what should be done in light of the technical considerations in Chapters 2 through 8 and the regulatory and other factors considered in Chapters 9 and 10.

11

What Should Be Done: INEEL HLW Calcine

Because the sodium-bearing waste (SBW) and the existing calcine are different classes of waste [mixed transuranic (TRU) low-level waste (LLW) and high-level waste (HLW), respectively] with differing characteristics, they should not be treated in the same way, processed in the same equipment, or disposed of at the same disposal site. Indeed, because disposal and transportation options differ for these two waste types, adopting the same treatment method, waste form, and disposal site is almost certainly not optimal. Therefore, treatment and disposal strategies for SBW and calcine are considered separately, with calcine discussed in this chapter and SBW in Chapter 12. Whether either of these wastes should be processed in the near future is also discussed, since the apparent risks associated with the current storage of SBW and the HLW calcine differ.[1]

The chapter considers what should be done with the Idaho National Engineering and Environmental Laboratory (INEEL) HLW calcine in light of

- constraints of regulations and cost (discussed in Chapter 9),
- as-yet undefined boundary conditions that impact the ultimate disposition of INEEL wastes (discussed in Chapters 9 and 10), and
- legitimate opportunities for future relief from milestones of a nonregulatory nature.

To expand on the third point, given that the year 2035 target deadline is not a *regulatory* constraint and is subject to renegotiations (in fact, the Settlement Agreement makes explicit provision for this), the committee, in addressing its Statement of Task, considered technical treatment options that would meet this target as well as options that would not. The result is shown below, as the committee position of what option *should* be chosen.

As discussed in Chapters 9 and 10, the committee believes that *no clearly acceptable pathway for ultimate disposal of the HLW calcine has been defined*, and therefore, more information than is currently available is needed to make an informed technical decision among the various processing options. The ultimate disposal location is uncertain, the form in which it would be acceptable at that location may differ from current specifications developed for the first repository, and the transportation pathway to get there (with its regulatory requirements)

---

[1] In addition to these difference in waste classifications and potential risk associated with current storage, SBW and HLW calcine are under different timeframes (with planned deadlines of 2012 and 2035, respectively) for remediation, as another reason why the committee has considered these waste streams separately.

is similarly an open question.[2] *The committee believes that aggressive efforts should continue to define these three points*, including the development of a regulatory solution for the Resource Conservation and Recovery Act (RCRA) constituents (e.g., delisting petitions, and confirmation that organics do not survive the calcining process, if this information is needed).

In the meantime, it is the committee's strong recommendation that the processing of calcine be deferred for a time limited to the duration within which the bins maintain their integrity, with modifications to the bins made as necessary to ensure safe storage. If and when during this period an acceptable pathway for ultimate disposition is established, as described above, an evaluation should then be made as to whether to pursue that route or continue storing the calcine in the bins.

If calcine were stored for a few decades, the concentrations of the radioactive species would be reduced to a level of activity within the regulatory limits for "remote-handled" (RH) TRU waste.[3] For longer time frames, the TRU activity would be reduced even further. For example, in approximately 500 years, the TRU activity would be reduced by approximately a factor of ten,[4] if calculations (Garcia, 1997) of initial inventories of the isotopes ^{238}Pu and ^{241}Am are correct, as could be confirmed by subsequent measurements. After approximately 500 years, the Cs and Sr activity would decay by about five orders of magnitude to approximately Class A LLW limits (see Figure 11.1). The consequence of this decay is that the residual activity (several hundred nCi/g of TRU elements, with additional activity from residual fission products) in the calcine, coming from fission products such as ^{99}Tc, ^{93}Zr, and ^{93m}Nb and actinides such as ^{239}Pu, ^{240}Pu, and ^{241}Am (Garcia, 1997: Tables 9 and 10), would represent a level of activity within the regulatory limits for "contact-handled" (CH) TRU waste (Garcia, 1997: Tables 9 and 10).

As these calculations show, the calcine could be processed with less technical risk and less potential radiological exposure if the processing is deferred beyond the near term. This processing could segregate selected species to generate a low-activity fraction for disposal on-site (meeting regulatory limits such as Class A LLW limits) and a smaller TRU fraction for disposal in a suitable repository. The radiation hazard associated with retrieval and processing of the calcine will lessen over time. According to Engineering Design File "EDF-BSC-002" of Dahlmeir et al. (1998), the radiation levels to be expected during calcine removal are in excess of tens of Roentgens (R) per hour, and possibly hundreds or thousands of R per hour.

The only near-term action should be improved confinement and stabilization of the bins, ventilation system, and associated equipment. The hygroscopic content of the calcine requires that there be adequate protection from atmospheric moisture if "caking" is to be avoided. A risk assessment is strongly recommended to identify which of these and other actions are most important to reduce risk. The risk assessment should also substantiate that this

---

[2] That is, it is not certain that the nation's first repository will be the one to contain the HLW calcine. Moreover, the waste acceptance criteria (WAC) for waste forms to qualify for disposal in the first repository are subject to future change, and the WAC for a second repository are unspecified (see discussion of these issues in Chapters 9 and 10). Because of these uncertain conditions, the full set of specifications on any treatment process to produce a final waste form cannot be determined at present.

[3] In support of this statement, the WIPP waste acceptance criteria for RH-TRU wastes specify a maximum radioactivity concentration of 23 Ci per liter (23,000 Ci per cubic meter), which Figure 11.1 shows is attainable for the HLW calcine after a few decades of decay after reprocessing.

[4] This factor of 10 is primarily due to the decay of the initial inventories of ^{238}Pu ($T_{1/2}$ = 88 years) and ^{241}Am ($T_{1/2}$ = 432 years). Some ^{241}Am is produced over time via beta decay of the (small) initial inventory of ^{241}Pu. Much of the HLW calcine originated from reprocessing of Navy SNF enriched in ^{235}U. While in reactors, successive neutron captures and beta (β) decays [specifically, ^{235}U(n,γ)^{236}U, ^{236}U(n,γ)^{237}U, ^{237}U(β$^-$)^{237}Np, ^{237}Np(n,γ)^{238}Np, and ^{238}Np(β$^-$)^{238}Pu)] produced the predominant amount of ^{238}Pu (another, much more minor source of ^{238}Pu is from alpha decay of ^{242}Cm), which decays with a relatively short half-life compared to other TRU isotopes. This significant fraction of ^{238}Pu in the waste means that INEEL calcine, more than other DOE HLW inventories, can be remediated via natural radioactive decay while in interim storage (Figure 11.1).

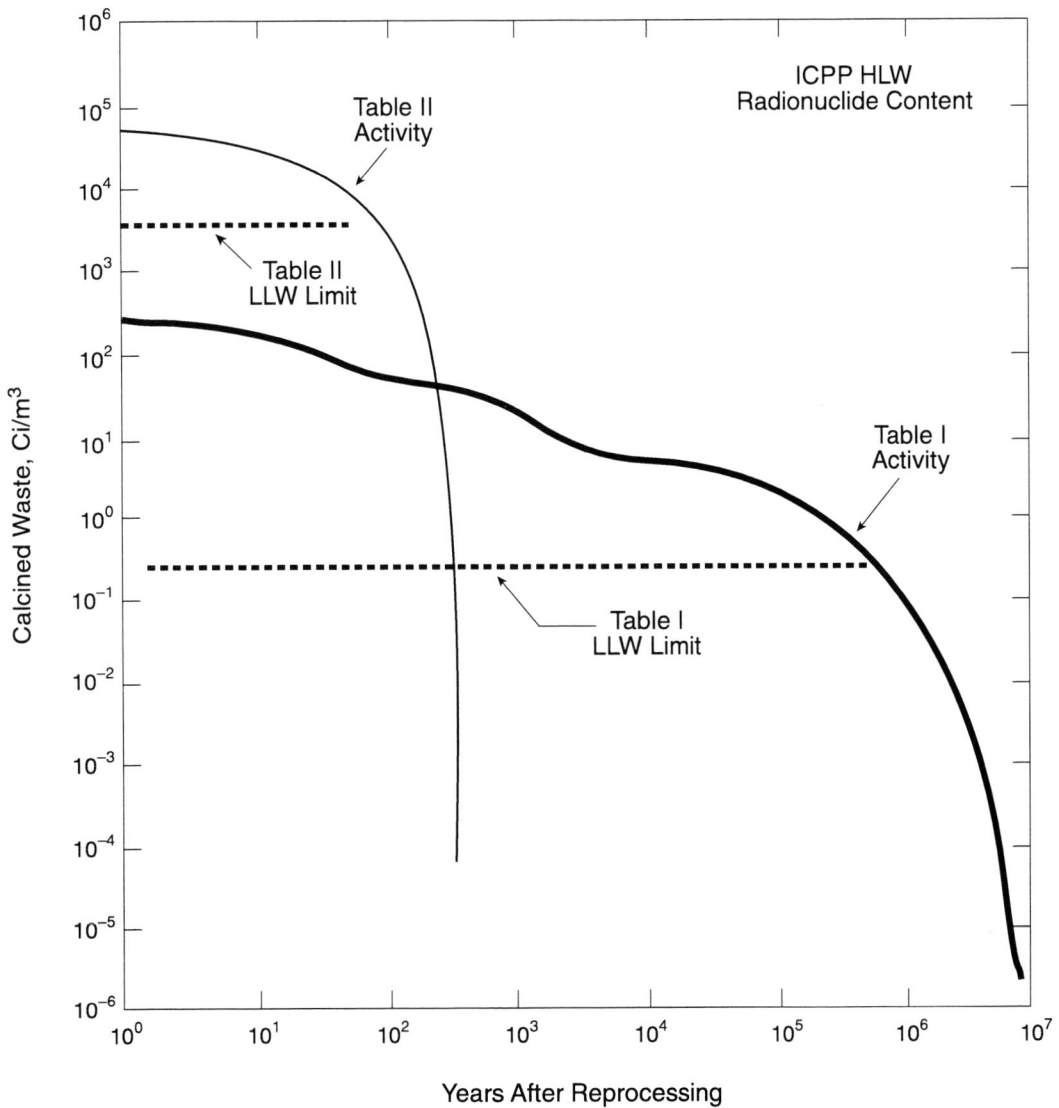

FIGURE 11.1 This graph shows decay in activity of INEEL HLW calcine over time. The longer-lived radionuclide constituents, which are actinides and long-lived β emitters listed in Table I of 10 CFR 61, are reduced by approximately a factor of 10 after 500 years. The shorter-lived radionuclides, which are the fission products Cs and Sr and other species listed in Table II of 10 CFR 61, are reduced by approximately five orders of magnitude in 500 years. Also shown are the Class C LLW limits that Tables I and II provide for these two categories of radionuclides. From Berreth (1988).

storage of a mixed HLW with significant TRU content in a near-surface environment (a storage which would probably require regulatory relief) poses a low risk at present and in the foreseeable future to workers, the public, and the environment. This recommendation does not challenge the general strategy of geologic disposal for HLW, which is an issue outside the scope of this study, but emphasizes that decisions on the ultimate fate of the INEEL HLW should be postponed pending the resolution of waste management issues noted above and pending the results of adequate risk analyses.[5]

In this recommendation to defer processing of HLW calcine until site(s), route(s), and waste form specifications are firmly established, *no time period is specified for the duration of interim bin storage*. Limitations to this time period could come from

(1) a technical assessment of bin integrity over time, and/or
(2) regulatory and other requirements.

To expand on (1), the information provided to the committee does not specify the failure mode (e.g., corrosion or seismic stability) that is the most limiting for bin integrity, and does not indicate whether 500 years signifies a mean time to failure or another design criterion. The committee recommends that during any period of interim bin storage, continuing verification of bin integrity is essential. To expand on (2), if the Licensing Requirements for the Independent Storage of Spent Nuclear Fuel and High-Level Radioactive Waste (10 CFR 72) were to apply to INEEL bin storage, then a regulatory license could be granted for up to 20-40 years, with any renewals for time periods beyond that contingent upon sufficient technical justification to satisfy the requirements for a license extension. Such matters of regulatory strategy were not examined in detail by the committee, and would require attention in the event that resolution of ultimate disposal site(s), route(s), and waste form specifications is not attained in the near future.

PERSPECTIVE ON OTHER HLW INVENTORIES

To expand upon this last point, HLW inventories being processed or involved in plans for processing at other DOE sites are in configurations much less stable than the INEEL HLW calcine. At the Hanford Reservation, for example, liquid HLW is contained in carbon steel tanks, many of which are single-shell tanks known to have leaked. At the Savannah River Site, the liquid HLW is stored in single- and double-shell carbon steel tanks that are located close to ground water. These conditions appear to pose a greater hazard or potential for release than solid calcine stored in stainless steel bins at INEEL. Since the current calcine configuration seems to be one of low risk, the need for action is correspondingly less, and a rush to select a long-term treatment option unwarranted.

---

[5] If bin storage of HLW calcine persists for more than a generation, then this recommendation does counter the general view (OECD, 1995) of having the present generation "dispose" of its own long-lived radioactive waste in geologic repositories; instead, for INEEL HLW calcine, the present generation would "manage" this waste. Such a bin storage strategy is consistent with the "stepwise implementation of plans for geological disposal" that might take place "over several decades" and is also consistent with the admission of "the possibility that other options could be developed at a later stage" that is expressly stated in OECD (1995: p. 9).

12

What Should Be Done: Sodium-Bearing Liquid Waste

Chapter 4 briefly discussed the proposed treatment options for sodium-bearing waste (SBW), and noted how classification and disposal restrictions were important considerations affecting the treatment method and final waste form. Guided by these ideas, the committee presents in this chapter its views of what should be done with the SBW, in a discussion similar to that presented in Chapter 11 for the HLW calcine.

Because of the potential for tank leakage, a study of options should be made promptly to identify ways to stabilize the liquid SBW into solid waste forms that are suitable for disposal or, if necessary, interim on-site storage pending disposal or further processing. In this effort, the committee's preferred general approach for treating SBW is to convert it to a solid, low-level, remote-handled transuranic (RH-TRU) waste to be shipped to a TRU repository designed to handle it [e.g., the Waste Isolation Pilot Plant (WIPP)[1]] or, if necessary, stored on-site. Treatment methods suitable for this purpose may also generate a non-TRU low-level waste (LLW) product that can be shipped to a LLW disposal site or, if necessary, stored or disposed on-site.

Five candidate SBW treatment methods are described below and are consistent with Department of Energy (DOE) program plans to empty liquid from all tanks by 2012. These methods are low-temperature (< 150 °C) and involve evaporation, simple chemical adjustments such as pH adjustment or addition of precipitants, and in some cases, solid-liquid separation. These methods do not involve calcination, vitrification, or complex multistage separation processes. The first four candidate methods involve evaporation and hydroxide precipitation; the fifth accomplishes TRU (and also much of the Sr) precipitation by sequential lanthanum fluoride precipitation.

In contrast, a sixth solidification treatment method, calcination in a repermitted calciner, is a thermal process producing a calcine waste form. If SBW is made into calcine, this calcine should *not* be mixed with the existing HLW calcine, principally because this changes its waste classification and thereby restricts eventual disposal options. While the calcination of

---

[1] WIPP's suitability as a TRU repository for these wastes would involve resolution of the issues raised in Chapter 9.

SBW and addition of the resulting calcine to bins of HLW calcine may be straightforward technically, this procedure creates additional inventory of HLW by mixing. HLW is the waste category for which final disposal options are most limited and restrictive.

SBW CONCEPTUAL FLOWSHEETS

Six examples of possible processes are given below. They vary as to (1) whether they develop two separate waste fractions, one a TRU waste and the other a non-TRU LLW; (2) how this separation is done, if at all; and (3) the relative amounts of the two fractions. DOE research and development efforts over the next few years should focus on developing sufficient information (much of which already exists at other DOE sites in the form of treatment options that can be adapted for use on the SBW) for a future decision to select a satisfactory flowsheet to process the SBW.

SBW Option 1: Acid-Side Direct Solidification

This option is conceptually the simplest. The SBW would be retrieved and evaporated to dryness or near-dryness (about 120 °C) such that the slurry concentrate would solidify when cooled. Water and nitric acid could be recovered from the overheads. The concentrate, containing all the radioactivity and salts in the SBW, would be solidified in drums or other appropriate containers and either shipped to a suitable TRU repository as RH-TRU waste (preferred) or stored on site. The evaporation step would probably be a two-stage operation, using a thermo-syphon first stage to concentrate the SBW and a wiped-film evaporator to carry it to a concentrated slurry product that would solidify on cooling.

The free-water content can be controlled by adjusting the temperature of evaporation (i.e., the boiling point). The slurry should be evaporated to the point that it contains less water than will be tied up as water of crystallization in the various salts present after cooling. As a result, there would be no free water. The slurry product would essentially be a salt cake and could be made into various physical forms, such as a monolithic salt cake or a powder.

The primary technical problem with this approach is probably evaporator and storage container corrosion with the acid system containing fluoride. If this is serious, one of the other options listed below should be used. Another disadvantage of this option is that all salts in the SBW go to the final waste, thereby generating a large volume of RH-TRU solid waste. The major advantage is that the only unit operation is evaporation.

SBW Option 2: Acid Destruction, Neutralization, And Direct Evaporation

The excess acid in SBW would be destroyed by addition of formic acid during an evaporation-concentration step with a conventional evaporator. The concentrate, containing little excess acid, would be neutralized with NaOH to a pH of about 8 to 10, precipitating most of the polyvalent metals (mostly Al) present. The slurry would then be evaporated and solidified for disposition as in Option 1. There has been considerable study of this method as well as other approaches at the Oak Ridge National Laboratory (ORNL) in managing neutralized waste reasonably similar in composition (McNeese et al., 1998).

The possible corrosion problem in Option 1 is due to evaporation of acidic fluoride solutions and would be largely circumvented in Option 2. Both of these options have the disadvantage that all the salts in the waste are converted to RH-TRU waste and must by disposed of in a suitable way.

SBW Option 3: Acid Destruction, Neutralization, Solid–Liquid Separation, and Solidification

As in Option 2, addition of formic acid, evaporation, and neutralization with NaOH would be done, producing a slurry. The slurry, after neutralization, would be treated for metals regulated by the Resource Conservation and Recovery Act (RCRA) using best available technology (probably sulfide addition), and then would go to a solid–liquid separation (SLS) step. The TRU, Al, other insoluble metal hydroxides, along with RCRA metals and much of the Sr would go to the solids fraction. The bulk of the salts (Cs, Na, K, and nitrates) would remain in solution. The solids fraction would be dehydrated, converted (e.g., by grouting) to a monolithic solid, if required to meet transportation or repository acceptance criteria, and shipped to a suitable TRU repository as RH-TRU waste or stored on-site. The liquid fraction would be evaporated, probably in two stages as in Option 1, with the wiped-film evaporator product being a solid non-TRU LLW that would be stored on-site or shipped to a suitable LLW repository as a Class C waste. There is considerable evidence from work at other sites with similar wastes that the supernate indeed would be non-TRU.

The primary disadvantage of Option 3 is the use of a somewhat difficult SLS process to recover a fairly large amount of precipitate. The advantage over Options 1 and 2 is that less TRU waste is generated.

SBW Option 4: Acid Destruction, Alkaline Leaching, SLS, and Solidification

This is similar to Option 3 except that more NaOH is added to the SBW to increase the hydroxide concentration enough to redissolve the Al. In this case, the SLS step is simpler because the amount of solids is smaller. Again, the solids contain the TRU and insoluble metal hydroxides including RCRA metals, and much of the Sr. The solution contains the soluble salts plus Al, and would be non-TRU LLW. The solids fraction would be evaporated and solidified to a form suitable for disposal in a TRU repository as RH-TRU waste, as in Option 3, but the quantity of solids would be smaller. The supernate would also be evaporated and solidified for storage or disposal, as in Option 3, but the quantity would be larger because of the Al and added NaOH.

The disadvantage here is that more NaOH is added, thereby increasing the amount of LLW. The advantage is that the TRU waste volume is smaller than for Option 3. Option 4 is analogous to the "enhanced sludge leaching" process at Hanford. The overall process is also quite similar to the method planned for disposition of a similar waste at ORNL under a recent privatization contract (Brass, 1998 ; DOE, 1998c).

SBW Option 5: TRU Separation by Lanthanum Fluoride Precipitation

The SBW would be processed in fluoride solution by addition of lanthanum. Multiple LaF_3 scavenges would precipitate a TRU fraction as an insoluble fluoride. The precipitate would be a high-density solid containing the rare earth elements and a large fraction of the Sr, and could be further processed, if necessary, into a more durable waste form. In this process, excess fluoride can be removed from the supernatant solution by adding sodium borate and boiling to drive off the excess fluoride as BF_3, if other fluoride complexants (e.g., Zr and Ca) do not interfere.[2] The lanthanide and actinide solubility product is so small that the SBW

---

[2] Precipitated plutonium fluoride is readily dissolved in boric acid (NRC, 1965: p.27). For example, it has been standard procedure at the Lawrence Livermore National Laboratory to concentrate the rare earth and actinide

could be converted to a non-TRU product with only a few LaF$_3$ scavenge cycles, assuming adequate mixing and precipitation at 85 °C and fast settling of the dense fluoride precipitate.

SBW Option 6: Conversion to Calcine and Storage Apart from HLW Calcine

If the calciner can be modified to satisfy current regulatory requirements, another solidification option is to calcine the SBW (with additives such as aluminum nitrate [Al(NO$_3$)$_3$], and store the resulting solid calcine in *separate*, RCRA-approved storage facilities. This option does not mix the new calcine with the HLW calcine. One disadvantage is that it requires operation of the calciner, and hence development work to repermit it to satisfy applicable U.S. Environmental Protection Agency regulations. Another disadvantage is that, compared to the other options, the solid form produced here is relatively harder to redissolve if future liquid-phase separations are needed.

SUMMARY

A more comprehensive examination of options such as 1 through 6 above is clearly required, based on a review of the extensive literature and relevant experimental work. At that point, a comparative evaluation could be used to rationally select the best approach. The choice would likely be influenced strongly by the availability of repositories that would accept the different waste products. If a suitable TRU repository is available, then options such as 4 and 5 above have appeal in that they provide relatively small RH-TRU fractions for disposal in this repository, albeit with added complexity in the processing steps. If a suitable TRU repository is not available for the TRU fraction of solids developed from the SBW, then it appears that at least the TRU fraction must remain on site. Nevertheless, the committee's position is to convert the liquid SBW into a solid form, preferably one that is very likely to be suitable for shipment to a future suitable TRU repository, if and when one becomes available.

This report recommends separate treatment of SBW and HLW calcine waste streams. This approach could be challenged. For example, if vitrification or cementation is to be performed, mixing the SBW (containing the sodium-fluxing agent) and HLW calcine (containing refractory elements) would serve to reduce the final waste volumes compared to the separate vitrification or cementation of the two waste streams. In response to this challenge, Chapters 4, 9, and 12 of this report note potential advantages (i.e., an opportunity to use non-HLW disposal options and more simplified treatment options) if the SBW is segregated from HLW and treated separately. These are benefits that can be assessed and quantified in the near term.

If the approach of this chapter is adopted, the final waste form for SBW need not be a vitrified one, and the SBW need not go to a HLW repository. Combining the SBW and HLW calcine would force the SBW to become classified as HLW via mixing, thereby restricting disposal options.[3] The committee's proposal to remove the SBW from the HLW stream achieves greater HLW volume reduction than the specific proposal in the preceding

---

fraction from dissolved Nevada fused rock (from Nevada Test Site tests) by precipitating a hydroxide precipitate from dissolver solution with ammonia, dissolving that in 2 to 3 molar (M) HCl, making the solution 2 M in fluoride, and then heating to 85 °C to agglutinate the fluoride precipitate. If performed correctly, this will quantitatively precipitate all actinides and lanthanides free of nearly all other elements. The fluoride precipitate can be readily redissolved in 2 M HCl (or HNO$_3$) by adding sodium borate and heating to 85 °C for only a few minutes with minor stirring. The fluorine is discharged as BF$_3$. On a large-scale operation, the BF$_3$ can be recovered as B$_2$O$_3$ from a water scrubber and recycled.

[3] This combination of waste streams might also introduce enough sodium into the HLW calcine to make it a complication.

paragraph.[4] Moreover, if the SBW were solidified as a saltcake and stored on site, this waste could be later added to the HLW calcine if this is ever deemed desirable.

---

[4] As discussed in Chapter 10, it is unknown at present whether HLW volume reduction is of overriding importance for a second repository. The cost consequences of these actions cannot be well quantified at present.

13

Summary of Conclusions and Recommendations

In a committee report of this type, with all of the information provided in the form of written reports and/or oral presentations, almost any observation or comment beyond a restatement of the material presented can be seen as a conclusion and/or recommendation. That is, a technical discussion of process steps leads naturally to committee statements of conclusions and recommendations. Chapters 2 through 12 contain many such statements. The reader is referred to them for a complete statement of committee views. What follows is a summary of the most important of these views.

WHAT SHOULD BE DONE: SODIUM-BEARING LIQUID WASTE (CHAPTERS 4 AND 12)

The committee believes that the sodium-bearing waste (SBW) [currently contained in tanks not approved under the Resource Conservation and Recovery Act (RCRA)] should be solidified in the near future with a process to be selected from a comparison study of proven methods that can be adapted for use on the SBW. The solid form produced should, if practical, be made acceptable for shipment to the Waste Isolation Pilot Plant (WIPP) or another suitable transuranic (TRU) waste repository. Any non-TRU low-level waste (LLW) fraction should be processed in a form acceptable for disposal in a LLW repository [e.g., the Nevada Test Site (NTS)]. Possible routes to accomplish this are described in Chapter 12.

If calcination of SBW (option 6 of Chapter 12) is done, the calcine product should be stored separately from the existing high-level waste (HLW) calcine inventory. Since SBW is not HLW, it is counterproductive to convert it to HLW (via mixing) or force it into a scenario designed for HLW because that decreases the options for processing and ultimate disposal (with a concomitant decrease in program flexibility).

WHAT SHOULD BE DONE: HLW CALCINE (CHAPTER 11)

The committee could identify no significant present hazard to public health or to the environment due to the storage of solid calcine in the bins at INEEL, which have been designed to be secure for at least 500 years. The need for immediate action and a rush to select a long-term treatment option appear unwarranted, in the committee's view, especially in comparison to significant inventories of HLW at other DOE sites that are in liquid form in underground tanks, some of which have leaked.

In contrast to the SBW storage situation, the committee recommends that no action to process the HLW be taken until it is clear where the material will be sent, what disposal form(s) is(are) acceptable, and that a viable transportation pathway is approved. The processing of calcine can be deferred so long as the bins maintain their integrity. This approach may require future modifications to the bins if necessary to ensure safe storage. In the meantime, the interim storage of HLW calcine in the bins is more practical than near-term treatment in view of several important considerations:

1. the radiation hazard associated with the near-term retrieval and processing of the calcine;
2. the integrity and long design life of the bin sets; and
3. the isotopic composition of the INEEL calcine, which differs from other major Department of Energy (DOE) HLW inventories in its relative abundance of transuranic isotopes that decay significantly over the timescale associated with the design life of the bin sets.

Additionally, interim bin storage of calcine is more practical than any other option because of the absence of an approved pathway to a different destination.

The only near-term action should be improved confinement and stabilization of the bins, ventilation system, and associated equipment. The hygroscopic content of the calcine requires that there be adequate protection from atmospheric moisture if "caking" is to be avoided. A risk assessment is strongly recommended to identify which of these and other actions are most important to reduce risk. The risk assessment should also substantiate that this interim storage of a mixed HLW with significant TRU content in a near-surface environment (a storage which would probably require regulatory relief) poses a low risk at present and in the foreseeable future to workers, the public, and the environment.

The committee believes that a risk analysis will show that extended interim bin storage of calcine is a low risk (and low cost) course of action. However, the committee emphasizes that this course should be subject to continued review and updating of comparative risks as additional information may be developed in the decades to come with respect to site availability, acceptable waste forms for such a site, and available transportation to such a site. This recommendation does not challenge the general strategy of geologic disposal for HLW, which is an issue outside the scope of this study, but emphasizes that decisions on the ultimate fate of the INEEL HLW should be postponed pending the resolution of waste management issues noted above and pending the results of adequate risk analyses.[1]

In this recommendation to defer processing of HLW calcine until site(s), route(s), and waste form specifications are firmly established, *no time period is specified for the duration of interim bin storage*. Limitations to this time period could come from

(1) a technical assessment of bin integrity over time, and/or
(2) regulatory and other requirements.

To expand on (1), the information provided to the committee does not specify the failure mode (e.g., corrosion or seismic stability) that is the most limiting for bin integrity, and does not indicate whether 500 years signifies a mean time to failure or another design criterion. The

---

[1] If bin storage of HLW calcine persists for more than a generation, then this recommendation does counter the general view (OECD, 1995) of having the present generation "dispose" of its own long-lived radioactive waste in geologic repositories; instead, for INEEL HLW calcine, the present generation would "manage" this waste. Such a bin storage strategy is consistent with the "stepwise implementation of plans for geological disposal" that might take place "over several decades" and is also consistent with the admission of "the possibility that other options could be developed at a later stage" that is expressly stated in OECD (1995: p. 9).

committee recommends that during any period of interim bin storage, continuing verification of bin integrity is essential. To expand on (2), if the Licensing Requirements for the Independent Storage of Spent Nuclear Fuel and High-Level Radioactive Waste (10 CFR 72) were to apply to INEEL bin storage, then a regulatory license could be granted for up to 20-40 years, with any renewals for time periods beyond that contingent upon sufficient technical justification to satisfy the requirements for a license extension. Such matters of regulatory strategy were not examined in detail by the committee, and would require attention in the event that resolution of ultimate disposal site(s), route(s), and waste form specifications is not attained in the near future.

As a final observation, the committee believes that, along with good science and engineering, a major consideration in deciding how (and *whether*) to process any radioactive waste for long-term conditioning is that of the risks being added and/or mitigated. The fundamental purpose of environmental regulations [such as those of RCRA, the Comprehensive Environmental Response, Compensation, and Liability Act (CERCLA), and United States Nuclear Regulatory Commission (USNRC) directives] and radioactive waste policy legislation must be, in the committee's view, minimization of risk to human health and the environment in a cost-effective and meaningful manner. A driving consideration in deciding upon a radioactive waste management strategy should be an identification, definition, and evaluation of the "trade-offs" (i.e., comparative risks) for the alternatives being considered, including those of limited or no processing. Such risk assessment calculations provide information on risk reduction strategies and are required to decide among alternatives in an informed and objective manner. Consequently, the committee believes that a risk analysis for the actions recommended above for both HLW calcine and SBW should be conducted promptly, and should include a comparison of the risks associated with INEEL HLW calcine and SBW to the risks associated with site inventories of other radioactive wastes. A sufficiently rigorous analysis should be performed to establish the current risks and to assess the changes in risk due to treatment options. In the committee's view, at least until the issues identified here are resolved, the risks (and costs) of repackaging the HLW calcine, with no certainty that it can ever be shipped outside of Idaho, may far exceed the risks (and costs) of continued interim bin storage.

CONCLUSIONS AND RECOMMENDATIONS
SPECIFIC TO INDIVIDUAL PROCESS STEPS

Conclusions and recommendations pertinent to specific steps of HLW processing options are given below. *They are of major importance only if the above recommendations with respect to SBW and calcine are not pursued.* Further details are provided in Chapters 2 through 10.

Calcine Characterization (Chapter 2)

The present database with actual aged calcine is inadequate to assure that processing goals can be achieved or to allow definition of processing methods without excessive technical risk. In addition, there appears to be no realistic sampling plan to add to this database by retrieving actual aged calcine, nor a characterization plan for such calcine. Actual aged calcine is crucial for testing purposes. Characterization of additional calcine samples from many bins is necessary to define retrieval and treatment issues adequately. The existing database for all work with actual calcine is inadequate as a foundation to design a treatment plant. Analytical data appear to be incorrect and inconsistent in some (too many) cases (see Chapter 2). This inaccuracy of analytical data is cause for concern because it calls into question not only the state of knowledge of calcine characteristics, but also the validity of projections based on the data. Adequate samples to characterize the waste, along with a sampling and analysis plan with appropriate quality controls, are necessary to collect essential data, determine properties such as solubility and dissolution of calcine in nitric acid, support development studies of treatment processes, and support risk analysis calculations. Most process development should be done with actual aged calcine.

Retrieval of Calcine from Bins (Chapter 2)

It is likely that calcine can be retrieved from the bins, but operational problems should be expected. The existing database is inadequate to define the problems that might arise and to evaluate how to resolve them. For example, the presence of hygroscopic constituents in the HLW calcine raises the issue of whether "caking" has occurred, which would prevent simple pneumatic transfers from being fully effective. A limited sampling and characterization plan would diminish uncertainties in retrieval and processing, thereby reducing the risk, and should receive high priority. The quantity of residual calcine allowed to remain in the bins after retrieval should be rationally defined. Consideration should be given to the handling of retrieved calcine with respect to providing stable and uniform feed for downstream processing (i.e., blending).

Calcine Dissolution (Chapter 2)

The proposed dissolution approach probably can be made to achieve the desired separations levels. However, the risk is substantial that the program will fail to achieve the separations goals (especially Class A requirements) without a substantial and sustained test program using actual aged calcine. The unknowns associated with the quantity and nature of radionuclides in undissolved solids (UDS) translate into uncertainty and risk in the design of a processing operation. The root cause of this concern is a lack of sufficient characterization and testing data with actual aged calcine.

Solid–Liquid Separation (Chapter 3)

Both physical and chemical characterization data for UDS from actual aged calcine are extremely limited and inadequate to define the solid–liquid separation (SLS) system, or even to show that SLS requirements can be met with a practical system. Such characterization data are necessary to establish requirements for the removal of solids and to identify and test promising approaches for achieving these requirements. Because entrained solids will degrade the attainable decontamination factor (DF) and jeopardize process operability in all downstream decontamination operations, efficient removal of particles down to very small sizes is required. Therefore, a key requirement for successful performance is the highly efficient removal of solids from the feed to any downstream chemical separations process (as estimated in Chapter 3, a factor greater than 10^5 is needed).

A reported study of the most promising SLS approach, single crossflow filtration, indicated limited effectiveness and problems caused by pore blockage. The apparent near-absence of a viable SLS program suggests that the necessity for effective SLS is not fully appreciated. The SLS program as presently conceived constitutes a region of unreasonably high technical risk.

In addition to the characterization program noted above, priority should be given to evaluating and testing at least two different methods for SLS. This work should be done with actual waste from SBW tanks and dissolved calcine rather than surrogates. Process residuals, secondary waste streams, emulsions, and problems with plugging are the types of concerns that should be resolved in such testing.

Cesium Ion Exchange (Chapter 3)

Three inorganic ion exchange materials have been tested for removal of Cs from both SBW and dissolved calcine under a limited range of conditions. The selected materials are reasonable but unproven candidates, and they are subject to uncertain future availability. Other promising sorbents have not been tested. Results to date are inconsistent and so limited that system performance cannot be predicted with a reasonable degree of confidence. The choice of sorbent has changed recently and cannot be considered to be well defined.

Evaluation has been limited to ion exchange column operation with elution and sorbent regeneration to allow repetitive cycles with a given batch of sorbent. It is questionable that the Cs DF required for Class A product can be achieved because, in subsequent loading cycles, residual Cs left on the column will slowly elute. If an elutable sorbent cannot be used to achieve sufficient decontamination of the waste stream, both the basis for selection of the sorbent and the mode of operation would be different, and final waste generation would likely be increased. The committee notes that Cs ion exchange is not required for a Class C waste product.

Extensive testing of promising Cs separations methods with actual waste solutions is required to demonstrate process viability. Particular problem areas relate to performance (DF and capacity) in subsequent cycles following elution, accumulation of solids or other constituents that do not elute, system operability with high volumetric throughputs, and secondary waste generation. Consideration of systems other than columns may be necessary.

Strontium Separation (Chapter 3)

The committee concurs with previous review groups that strontium extraction (SREX) is a promising approach for removing ^{90}Sr from SBW and dissolved calcine. However, the committee also believes that a number of significant technical difficulties described in Chapter 3 make achievement of Class A separations unacceptably risky without substantial additional chemical process information.

In agreement with the feasibility study (Fluor Daniel, Inc., 1997), the committee recommends sustained tests of the complete extraction, strip, wash, and acid-rinse subsystems. Critical process vulnerabilities to be investigated should include solvent extraction recycle and degradation, impurity buildup in the organic phase, temperature effects, and formation of precipitates and emulsion. In particular, the behavior of RCRA constituents such as Pb and Hg needs to be clarified in tests with a range of actual aged calcines, because of known difficulties in stripping Pb and Hg from the organic phase and the formation of interfacial Pb precipitates.[2] Additional process testing is needed to solve these difficulties in order to demonstrate the capability to meet Class A separations requirements, which are extraordinarily demanding and therefore likely to require significant resources.

---

[2] The Pb and Hg constituents also pose interference problems for TRU separations.

TRUEX Separations (Chapter 3)

TRU separations have been proposed by using the TRUEX solvent extraction process applied to a feed solution consisting of calcine dissolved in acid. However, several of the nonradioactive chemical constituents of INEEL HLW calcines will complicate the successful extraction of the relatively small mass of TRU elements present. The chemical process design will depend on the chemical composition of a TRUEX feed solution and its inherent variability, as derived from the variations in calcine compositions upon retrieval and after any subsequent blending operations.

In particular, the HLW calcines contain significant quantities of chemicals (e.g., Zr and Fe) that may directly compete with the actinides in interactions with TRUEX chelating agents (e.g., CMPO; see Appendix F). The few laboratory-scale extractions that have been performed to date on actual aged calcine confirm this suspected interference but provide no practical alternative chemistry to solve this problem. Therefore, TRUEX processing may not work effectively or operationally for the full range of feed compositions without major head-end chemical steps to remove interfering species. If the range of chemical constituents in redissolved calcine solutions warrants the use of complex head-end treatments, both the complexity and operational costs of processing would increase significantly. In this event, a comprehensive reexamination of the TRU separation technology should be considered.

If significant TRU separations are required, then sufficiently large-scale demonstrations are necessary using actual aged calcine to test for the operability of a process under flowsheet conditions. These tests are particularly important in view of limitations on the present knowledge of (and control on) the variability in the chemical composition of the feed. Several of the nonradioactive species present in dissolved calcine can chemically interfere with the extractant to degrade the level of separation that is achieved. Also, solid residues (typically fluoride and phosphate precipitates) that can be formed during the back-extraction step can physically damage, or even destroy, centrifugal contactor cascades and thus force total plant failures. In view of these problems, the operability and efficacy of a large-scale TRUEX process under currently proposed flowsheets using dissolved calcine solution should be critically assessed prior to the commitment of significant resources. This "scale-up" assessment cannot be done from experiments and characterization data obtained to date.

This current status of INEEL TRUEX development efforts indicates that it would be premature to propose TRUEX partitioning to remove actinides from redissolved INEEL calcine to meet Class A limits in a large-scale process. The process development required to extract actinides from the chemically diverse feeds that can be formed from retrieved calcine is, in the committee's opinion, a challenge that is properly characterized as first-of-a-kind chemical processing design work. Therefore, an adequately funded chemistry investigation is needed to resolve outstanding issues. In this effort, nuclear and radiochemical processing expertise within the DOE and national laboratory systems and in private, including foreign, companies should be engaged more extensively.

General Conclusions on Chemical Separations (Chapter 3)

Challenges have been noted above for cesium, strontium, and actinide separations in a large-scale process designed to produce from the calcine a low-activity waste (LAW) stream meeting Class A limits. Before committing to large-scale SREX and TRUEX processing to meet Class A separations requirements, the following issues should be resolved:

1. *Effects during processing and final disposition of mixed waste species.* Mixed waste constituents, particularly the RCRA metals Hg and Pb, complicate both SREX and

TRUEX operations, and potentially cause both the high-activity waste (HAW) and LAW streams to be classified as mixed waste.

2. *Testing with actual aged calcine.* Additional testing with a sufficient range of actual aged calcines is necessary to confirm the limited results obtained to date with surrogates. These tests should include reagent recycle and sequencing of unit operations using actual solutions, and should be long enough in duration to demonstrate long-term operability.

3. *Scale-up demonstrations.* A successful pilot-scale demonstration of the proposed chemical separations process should be conducted prior to the commitment of resources to a full-scale system. Since downstream unit operations can be negatively impacted as a result of even minor changes in upstream operations, the testing should use actual aged calcine, full sequencing of operations, and reasonable run duration.

Vitrification (Chapter 5)

Among various solidification options, vitrification has been investigated extensively at INEEL and elsewhere. Frit glass compositions and simulated waste glasses have been developed for different calcine compositions. In general, the chemical durability of these glasses has been found to be satisfactory. The major effort remaining to complete this work is in the vitrification of the HLW resulting from the separation option, for which research is under way to develop an appropriate waste glass composition. In this research effort, a wider range of glass compositions, including phosphate glasses, should be explored. Vitrification facility plans have been made, and in some cases pilot-scale melting has been conducted. Experience at other DOE sites and laboratories may provide some useful guidance for these efforts.

The composition of calcine waste feed to a melter would depend on the retrieval and blending strategy. Methods to mix calcine recovered from several bins should minimize the variation in the overall composition of the blend. A large supply (perhaps a 1-month to 1-year supply) of blended feed might be necessary to ensure stable, continuous production of a vitrified waste form. Since it will be necessary to adjust the glass frit composition to compensate for variations in the blended calcine composition, a glass frit compositional effort should be planned to continue during the entire vitrification period.

Alternatives to a Borosilicate Vitrification in a Continuous Melter (Chapters 5 and 7)

A concern of the committee is whether borosilicate vitrification in a continuous melter is the best solidification option for the INEEL calcines. Although INEEL calcines are, like Hanford and Savannah River tank waste, classified as high-level, they differ in important ways. The INEEL calcines may be distinguished from these other HLW inventories by their relatively higher content of Ca, Al, Zr, and F and relatively lower content of Na. The high alumina and zirconia content of the calcine is expected to result in a low waste loading in a borosilicate glass type because of the low solubility of these oxides at the planned melting temperature of 1150 °C.

In view of the unique compositions of some of the wastes at INEEL and the loading problems they pose in borosilicate glass formulations, other solidification options, such as glass-ceramics, nonborosilicate glasses, and single-use melters, may be more advantageous by offering a satisfactory final form at lower cost or with higher waste loading or easier processing. In particular, a single-use melter offers the possibility of very high waste loadings while avoiding many of the problems inherent to continuous melters. Immobilization options (e.g., phosphate glass production, or partial vitrification; see Chapter 7 for a more complete list) other than borosilicate glass production also appear to have merit and therefore should be considered. In practice, this means that these other options should be included in a comparison of options, as in the Environmental Impact Statement (EIS) analysis, especially if the subset of options in the EIS involve extensive aqueous processing prior to vitrification and/or a large volume of HLW glass.

Cementitious Options (Chapter 6)

Cementation processes can be used to produce a HLW grout as well as a LLW grout. The cementation options proposed in DOE literature are all quite similar insofar as unit operations and equipment are concerned. All involve mixing of dry materials with liquid and subsequent placement of the semi-liquid material in containers. The major differentiation between the options is what has occurred to the waste prior to mixing with cement and what waste form results from the constituents after cementation.

Any of the cementation options could be made to work easily and with reasonably low equipment costs. The direct cementation options (i.e., without chemical processing steps prior to waste immobilization), as compared to the options that contain processing steps, contain less risk and fewer personnel exposure hazards and produce lower overall waste volumes, albeit the volume of high-activity waste (HAW) may be much larger.

There would still be a significant amount of work to be done to develop a cementation process. Representative samples of the various calcines must be obtained to characterize their composition and quantity of each type. Sampling of waste tanks of SBW is necessary for the same reason. Knowledge of compositions is required so that appropriate recipes for cementation can be developed. A pilot line (not necessarily full scale) needs to be set up and run first with simulated compositions (cold runs) and then installed in a hot cell for demonstration runs to prove processing feasibility and to produce samples for testing. To fully specify process requirements, work on resolving issues of RCRA constituents (e.g., delisting) should be initiated immediately. Similar development work would be needed for any treatment process, to test its viability on a sufficiently large scale on actual aged calcine.

There are inadequate data to support the contention that cementitious waste forms can be made more quickly, more cheaply, more simply, and more safely than other (e.g., vitrified) waste forms. That contention may be true, but it should be established by thoughtfully designed experimental programs and an analysis of comparable data (a full-scale demonstration of the process is not required). Although cement-based processes applied to INEEL HLW may prove to be relatively straightforward, the quantity of final waste that is acceptable to produce and the qualification of the cementitious waste form for suitable disposal are the major issues that would need to be resolved to make cementation a fully viable option.

Other Alternative Waste Forms (Chapter 7)

The committee has been presented with a number of alternative waste forms that could eliminate the need to redissolve the already solidified waste (calcine). Two important criteria for evaluating any type of waste form are (1) the ease of producing it with minimum exposure to workers, and (2) its long-term physical integrity and chemical durability which reduces the exposure to persons in the future. Unfortunately, only limited substantive data were presented to the committee for each of these waste forms, and in some cases, such as the simulated spent fuel (SSF), no data are available at this time.

The limited data for these alternative waste forms indicates a lack of current support for exploring these possibilities, with several documents given to the committee dating back to the early 1980s (e.g., Post, 1981). Much of the data needed to compare and analyze alternative waste forms is available in the published literature; therefore, expensive experimental programs are not necessarily needed to provide at least some of the missing data. As an example, Table 7.1 compares the properties of a sintered glass waste form prepared at the University of New Mexico to reported literature on partially vitrified waste forms prepared at INEEL. Similar compilations of comparative data could and should be made for each alternative waste form that is considered to have a reasonable potential for practical use.

Summary of HLW Immobilization Options

This report identifies attractive features of several immobilization methods, including: (1) SSF, (2) cementation, (3) phosphate-based glasses, (4) a glass/ceramic (specifically, those made by cold or hot pressing or by embedding calcine in a glass shroud made by "partial vitrification"), and (5) options other than continuous Joule-heated vitrification melters (e.g., a single-use melter). Each of these methods has some merit, with potentially attractive features as well as uncertainties and potential problems that have been identified in Chapters 5-7. From these chapters, the committee concludes that many waste forms and processes for their creation can be identified. For example, Chapters 6 and 7 cite cementation and SSF as probably less costly processes than direct vitrification for the HLW calcine, but they produce final waste forms that would require further development and proper qualification for acceptance for disposal. None of these immobilization proposals logically can be selected until the definitive criteria, set in part by the approved disposal pathway, have been established.

As a further consideration, the committee has identified the following features as a useful basis to compare the different immobilization methods and to gauge their challenges and likelihood of success. These other features are:

- waste form properties, including durability;
- waste loading and net material additions;[3]
- relative maturity and technical risk;
- operability and processing ease;
- worker and public risks;
- environmental releases and risks;
- regulatory requirements and challenges;
- estimated relative cost; unit size;[4]
- other controls on the process environment,[5]
- production experience.[6]

---

[3] This is the net amount of material added in pretreatment, processing, and/or immobilization steps. In general, to minimize final waste volumes, the least amount of net material should be used. The net material would include consideration of recycled materials that are recovered and reused.

[4] This is the repeatable unit that defines the process conditions and properties, which in turn impact issues such as safety and throughput. The unit size would determine the amount of calcine that is retrieved, characterized, and processed at one time, to include consideration of sufficient blending for feed homogeneity. In principle, there can be differences between the unit size used in research and developmental testing, the unit size used for full-scale production, and the unit size used for qualification of the waste form (e.g., the whole production unit could undergo qualification/validation, or just a test sample). The criterion useful in evaluating immobilization options is whether the scale set by the unit size is adequate (for testing or for throughput). In particular, the test unit size must be large enough to provide a basis for scale-up to production size with considerable confidence.

[5] Apart from environmental and safety controls that would be needed, other process conditions, such as the redox environment, may need controls that should be considered if they pose a technical challenge or limitation.

For example, in relative maturity, the technologies discussed in Chapters 2-7 might be assessed as shown in Table 13.1. The relative maturity is, as shown above, only one feature among many valid technical considerations that are useful to consider in an inter-comparison of options.

Tank and Bin Set Closure Issues (Chapter 8)

Closure operations for the tanks and bin sets will require inventive engineering solutions that are practical and feasible with current technology. However, clean closure of the tanks and bin sets is impractical, insofar as some residual radioactivity, and possibly hazardous chemical constituents, will remain in the facilities.

The options for closure of the tank farm and bin sets that are proposed in DOE literature are all possible from a technical standpoint. The health and safety of the public and the personnel performing the closure process is undefined pending agreements on acceptance criteria and the means for calculating risk. Assuming rational judgment and prudent management, it would be reasonable to expect that all options will result in very low risk.

Based on its review, the committee believes that the technical success of the closure process would be enhanced by the following initiatives:

1. Develop and innovate sampling strategies that focus on needs derived from a risk analysis of closure. If a risk-based closure is the operational driver for regulatory agreements, then sampling and analyses are needed to make this risk assessment sufficiently credible. For the risk information required, it may be insufficient to rely on sampling and analyses done for other purposes (such as process control of future separations of HLW, LLW, and hazardous chemicals and material balances of process schemes).

2. Consider the possible advantages of restructuring the calcine retrieval and transport project (CRTP) and bin set closure project (BSCP) programs into three functional responsibilities: (a) movement of calcine internal to the bin and piping (retrieval and decontamination), (b) movement of objects and facilities external to the bins (containment assurance), and (c) movement of material into the bins (stabilization). The present programs have potentially too many interfaces that may hinder the effective coordination of efforts.

3. Conduct essential testing such as proof-of principle tests and mock-up testing, and qualify key remote processes such as visual inspection of the tanks and bins, maneuverability, and bottoms management. Tank decontamination, which is scheduled first, could be used as a test-bed to qualify features to be used for bin decontamination. Also, the test program should anticipate process failure and develop and test recovery measures.

The closure program plans reviewed by the committee seem to be related to decisions based on risk analyses that are not independent of other Idaho Chemical Processing Plant (ICPP) and INEEL facility closures. Filling the tanks and bins with hazardous or LLW waste, storing road-ready HLW canisters, retaining the calcine in the bins, removing the tanks and bins by clean closure, and discarding them at another location on the site all contribute to the risk equation in the postclosure period. Each of these tasks has components of risk that should be considered in a comprehensive assessment.

---

[6] Relevant expertise could be derived at least in part from users of similar equipment and related technology. These sources might need to be tapped to surmount challenges, particularly those unforeseen initially. For example, the SSF concept is similar to commercial nuclear fuel fabrication technology, which provides somewhat relevant production experience.

Table 13.1 Qualitative Estimations of Relative Maturity of Various Technologies, Some Proposed by DOE and Some by the Committee, that are Useful in HLW Calcine Treatment

Process	Relative Maturity
Bulk Calcine Retrieval Techniques	Not proven technology but likely to work.
Retrieval of Smaller Samples	Proven technology (demonstrated in 1979).
Dissolution of Calcine	Feasible but uncertainties remain due to lack of extensive testing on actual aged calcine.
Solid-Liquid Separations (SLS)	Techniques are available but have not been integrated into a system because the system requirements are not yet fully defined. The SLS system requires substantial development and testing.
Cs, Sr, and TRU Separations	Tested on lab scale, mostly with simulants; limited testing with actual aged calcine. The application on INEEL HLW calcine poses problems (e.g., with constituents that interfere with separations processes) that require significant developmental testing, scale-up, and process integration.
Cementation	Feasible technology for HLW but not yet proven widely for HLW in general and INEEL HLW in particular.
SSF	Not yet demonstrated on HLW; development and testing needed to validate the concept.
Continuous Melter - Borosilicate Glass	Proven technology on some HLW, but formulation must be developed for the INEEL HLW compositions high in Zr, Al, and/or CaF_2.
Phosphate Glass	Proven technology on some HLW, but further development and testing needed on INEEL HLW.
Single-Use Melter	Demonstrated technology for radioactive waste but further development and testing needed on INEEL HLW.
Glass/Ceramic (Via Hot Isostatic Pressing)	Proven technology on INEEL HLW (but safety and throughput concerns provide limitations).
Glass/Ceramic (Via Cold Pressing followed by Sintering or Hot Uniaxial Pressing)	Demonstrated technology for radioactive waste, but further development and testing needed on INEEL HLW.
Partial Vitrification (Via Embedding Calcine)	Demonstrated technology for radioactive waste, but further development and testing needed on INEEL HLW.
Common to all Immobilization Processes	Qualification of waste form, regulatory approval for process and product(s), and large-scale testing of the process(es) are needed to categorize the technology as fully mature and implementable on INEEL HLW calcine.

To expand on this issue, residuals of SBW and HLW calcine will remain at the INEEL site regardless of any option considered to process these wastes. For SBW, a portion of sludge on the tank floor may be unrecoverable and therefore destined for in-situ grouting and burial. A fraction of the calcine inventory may also be unrecoverable during retrieval and decontamination practices, and similarly destined to be grouted in place. The types and quantities of these wastes remaining on site can be analyzed in performance assessment calculations as defining the end state of remediation (NRC, 1999; Appendix B).

The recovery of residual waste, and the amounts left in place, should be assessed in various scenarios to determine relevant "trade-offs" such as the cost benefits and risk reductions that are achieved. These assessments of closure options should be used to guide decisions of tank closure criteria and specifications.

Establishing a Disposal Path (Chapter 9)

The viability of any out-of-state transportation and disposal option for INEEL HLW calcine (or for waste products made from HLW calcine) is highly uncertain at present. Therefore, the program cannot rationally select a processing option based on a currently established disposition pathway for the waste products. Until such a pathway is defined, the calcine should not be processed to yield a product that is less amenable to further processing than the current calcine.

To expand upon the present state of uncertainty for HLW repositories, many aspects of the planned Yucca Mountain repository are uncertain and subject to change, including (1) whether DOE will be able to proceed with its construction, (2) what strategy will be followed to emplace waste, (3) how repository capacity will be allocated among various types of defense waste, (4) whether the current legal limitation of 70,000 MTHM will be changed by Congress in future legislation, and (5) waste acceptance criteria, which may be expanded and revised. The "second repository program" is even more uncertain and subject to change. The INEEL HLW could be relegated to this second repository program, insofar as the Yucca Mountain repository, if built, is slated to be filled by 2035. The waste form performance and waste acceptance criteria for the second repository program are not yet known, and may differ from those of the first repository, if indeed a second repository is mandated.

Waste Management Strategy (Chapter 10)

To meaningfully compare and decide among processing options that produce different waste forms, more information than is currently available is needed on (1) viable disposition pathways for waste streams and the requirements they impose on the waste forms, and (2) characterization of the calcine and SBW. Each of these inputs is needed to establish a credible basis for an informed choice between different proposed processing technologies and different waste forms (e.g., glass, cement, or a glass/ceramic).

In order to properly focus on where the HLW is to be sent, an overall waste management plan is needed. To be truly comprehensive, such a plan would encompass all site inventories of radioactive wastes and other nuclear materials. A key element of such a plan is a firm knowledge of where the material is to go and what form it would be in to be acceptable. Until those issues are resolved, the selection of any technical option is without a proper basis.

One purpose of a waste management plan is to treat uncertainties that prevent the full resolution of the technical issues in Chapters 2-8, the regulatory issues in Chapter 9, and the other program planning considerations in Chapter 10. The DOE planning approach in the past has been to make explicit planning assumptions [e.g., Wichmann et al. (1996) makes the assumptions that Federal environmental policies, strategic policies, and national priorities will remain unchanged; that treated INEEL HLW will be sent to the second (currently hypothetical) HLW geologic repository after 2065; that INEEL waste management activities may be licensed by other agencies, such as the USNRC; that Federal and State of Idaho environmental policies, statutes, codes and orders will remain unchanged; and that Congress will provide the funding necessary to meet the DOE's agreements with the State of Idaho]. The committee views these assumptions as an inadequate way to resolve uncertainties, and therefore proposes that the disposal conditions be firmly established prior to the selection of a treatment option. The committee recommends aggressive efforts to establish these disposal conditions. However, in practice, for the current situation, given that Yucca Mountain is to be filled by 2035 and a second repository program may not be identified for several decades, the firm establishment of disposal conditions may not be forthcoming in the near future, in which case following the recommendation of this report may translate into a suspension of effort in the near term to process the calcine into a different waste form.

At this stage in the process, acquiring data to replace or validate key assumptions is a more informative and useful activity than conducting design studies and cost estimates that are based on assumptions yet to be validated. In addition to addressing uncertainties, other important features of a waste management plan, in the committee's view, are to adopt approaches that achieve risk and cost reductions and that support the thoughtful use of science and engineering in developing disposal strategies and waste forms.

CHAPTER SUMMARY

Table 13.2 summarizes some of the major recommendations of this chapter. As noted previously, these conclusions and recommendations in Chapters 2 through 10 are of major importance only if the conclusions and recommendations regarding Chapters 11 through 12 at the beginning of this chapter are not followed. These recommendations from Chapters 11 through 12 are, in the committee's view, feasible and readily implementable. In contrast, the recommended activities from Chapters 2 through 10 involve a pursuit of not only separations and immobilization challenges associated with processing the calcine, but also regulatory and other challenges. In the committee's view, working through the various challenges of this latter course of action is a formidable task that would likely require resources external to the INEEL HLW program. The alternatives for calcine treatment have substantial uncertainties and as yet unresolved technical difficulties.

Table 13.2 Summaries of Major Recommendations

Component or Topic	Recommendation of this report
Calcine	Risk assessments should be performed to assess bin integrity and on- and off-site risks of continued interim bin storage and processing and disposal options. Processing should be deferred until: (1) the final repository location, disposal form, and transportation parameters are established; (2) an appropriate treatment method is identified, based on (a) sufficient characterization data of and testing with actual aged calcine, and (b) pilot plant testing at a realistic scale; and (3) further risk assessments confirm that such treatment constitutes less of a risk to human health or the environment than continued bin storage.
SBW	Identify suitable non-calcination solidification methods and select one to solidify this waste in the near term as a high priority.
Tank Closure	Establish closure criteria; in this effort, consider using a risk basis. Determine the data needs and develop an appropriate sampling and analysis plan.
Waste Management Plan	A plan should be developed to resolve the uncertainties identified in this report. These uncertainties are of a technical, regulatory, other legal, and/or DOE policy nature. The trade-offs among various alternatives should be exhibited and used in decision making.

If the above recommendations are not followed, then the recommendations below apply.

Characterization	Adequately sample and characterize the calcine and SBW to reduce uncertainties in retrieval, dissolution, SLS, and other downstream processes.
Calcine Dissolution	Testing on actual aged calcine is recommended to reduce uncertainties and risks.
Solid-Liquid Separations	SLS system requirements should be developed, after present unknowns and uncertainties in waste characterization and separations requirements are resolved. Testing with actual waste is needed to provide base information.
Cs, Sr, and TRU Separations	Further testing is needed, particularly on actual aged calcine and on the behavior of potentially interfering species in sufficiently large-scale demonstrations.
Solidification Options for HLW Calcine:	
Vitrification	A wider range of options should be explored, particularly alternatives to borosilicate vitrification in a continuous melter. Such alternatives include single-use melters and phosphate-based glasses.
Cementation	Further developmental work is needed.
Other Immobilization Options	Further developmental work is needed on methods such as "Simulated spent fuel" and glass ceramics produced by cold pressing and sintering, hot uniaxial pressing, or partial vitrification, all of which methods at this stage have merit and should be considered further.
Disposal Requirements	Establish disposal requirements in order to know (1) whether separations processing to reduce HLW volume is appropriate, and (2) which HLW immobilization process (and waste form) to select.

References

Bell, J. T. 1998. Alternatives to HLW Vitrification: The Need for Common Sense. Presentation given at SPECTRUM '98 September 13-18, in Denver, Colorado. Bell Consultants.

Berreth, J. R. 1988. Inventories and Properties of ICPP Calcined High-Level Waste. (WINCO 1050; February). Idaho Falls, Idaho: U.S. DOE/Westinghouse Idaho Nuclear Company, Inc.

Bonano, E. J., M. S. Y. Chu, L. L. Price, S. H. Conrad, and P. T. Dickman. 1991. The Disposal of Orphan Wastes Using the Greater Confinement Disposal Concept (DE-Ac04-76DP00789). Sandia National Laboratories Albuquerque, NM: U.S. Department of Energy.

Bonniaud, R., P. Labe, and C. Sombret. 1968. Vitrification des residus solides fluores provenant de l'attaque par le fluor des combustibles irradies. Presentation given at the Symposium sur le Retraitement par Voie Seche on October 29 in Mol, Belgium.

Brass, L. 1998. Foster Wheeler discusses project. Oak Ridge, Tenn.: *The Oak Ridger*. October 15.

Brewer, K. N., R. S. Herbst, T. J. Tranter, A. L. Olson, T. A. Todd, I. D. Goodwin, and C. W. Lundholm. 1995. Dissolution of two NWCF calcines: extent of dissolution and characterization of undissolved solids. Idaho National Engineering Laboratory (INEL-95/0098; February). Idaho Falls, Idaho: U.S. Department of Energy/Lockheed Martin Idaho Technologies Company.

Brewer, K. N., T. A. Todd, A. L. Olson, D. J. Wood, V. M. Gelis, E. A. Kozlitin, and V. V. Milyutin. 1996. Cesium removal from radioactive idaho chemical processing plant acidic waste with potassium copper hexacyanoferrate (INEL-96/0356; November). Idaho Falls, Idaho: U.S. Department of Energy/Lockheed Martin Idaho Technologies Company.

Brewer, K. N., R. D. Tillotson, P. A. Tullock, T. G. Garn, R. S. Herbst, J. D. Law, T. A. Todd, and A. L. Olson. 1997. Elimination of Phosphate and Zirconium in the High-Activity Fraction Resulting from TRUEX Partitioning of ICPP Zirconium Calcines. Idaho National Engineering Laboratory (INEEL/EXT-97-00836; July). Idaho Falls, Idaho: U.S. Department of Energy/Lockheed Martin Idaho Technologies Company.

Brewer, K. N., A. L. Olson, W. S. Roesener, and J. L. Tonso. 1997. Experimental results of calcine dissolution studies performed during FY-94, 95 (INEL-97-01192; September). Idaho Falls, Idaho: U.S. Department of Energy/Lockheed Martin Idaho Technologies Company.

Brotzman J. R. 1978. Vitrification of high-level alumina nuclear waste. P. 215 in Scientific Basis for Nuclear Waste Management II, C. J. M. Northrup, Jr., ed. New York: Plenum Press.

Campbell, D. O., and D. D. Lee. 1991. Treatment Options and Flowsheets for ORNL Low-Level Liquid Waste Supernate (December - ORNL/TM-11800). Oak Ridge, Tenn.: Oak Ridge National Laboratory.

Campbell, D. O., D. D. Lee, and T. A. Dillow. 1991. Development Studies for the Treatment of ORNL Low-Level Liquid Waste (November - ORNL/TM-11798). Oak Ridge, Tenn.: Oak Ridge National Laboratory.

Chen, F., and D. E. Day. 1999. Corrosion of Selected Refractories by Iron Phosphate Melts, *Ceramic Transactions*, 93:213-218.

Cochran, J., B. Crowe, and F. DiSanza. 1999. Intermediate depth burial of classified transuranic wastes in arid alluvium (abstract). Albuquerque, N.Mex.: Sandia National Laboratories.

Cole, H. S., and E. L. Colton. 1982. Comparison of candidate glasses to vitrify ICPP calcined alumina waste. *Transactions of the American Nuclear Society* 43:109-110.

Dafoe, R. E., and S. J. Losinski. 1998. Direct Cementitious Waste Option Study Report (INEEL/EXT-97-01399; February). Idaho Falls, Idaho: U.S. Department of Energy/Lockheed Martin Idaho Technologies Company.

Dahlmeir, M. M., L. C. Tuott, B. C. Spaulding, S. P. Swanson, K. D. McAllister, K. C. Decoria, S. L. Coward, R. J. Turk, R. A. Adams, and C. W. Barnes. 1998. Calcined Solids Storage Facility Closure Study (INEEL/EXT-97-01396 February). Idaho Falls, Idaho: U.S. Department of Energy/Lockheed Martin Idaho Technologies Company.

Day, D. E., Z. Wu, C. S. Ray, and P. Hrma. 1999. Chemically durable iron phosphate glass wasteforms. *Journal of Non-Crystalline Solids* 241:1-12.

Dirk, W. J. 1994. Long-Term Laboratory Corrosion Monitoring of Calcine Bin Set Materials Exposed to Zirconia Calcine. Idaho Falls, Idaho: U.S. Department of Energy/Westinghouse Idaho Nuclear Company, Inc.

Dole, L. R., G. C. Rogers, M. T. Morgan, D. P. Stinton, J. H. Kessler, S. M. Robinson, and J. G. Moore. 1983. Cement-Based Radioactive Waste Hosts Formed under Elevated Temperatures and Pressures (Fuetap Concretes) for Savannah River Plant High-Level Defense Waste (ORNL/TM-8579; March). Oak Ridge, TN: Oak Ridge National Laboratory.

Fluor Daniel, Inc. 1997. Idaho Chemical Processing Plant Waste Treatment Facilities Feasibility Study Report, Volumes I & II. Idaho Falls, Idaho: Department of Energy-Idaho Field Office.

Fluor Daniel Hanford, Inc. 1998. Hanford Site Solid Waste Acceptance Criteria for June 29 (Document No. HNF-EP-0063). Available at URL: http//www.hanford.gov.

Gahlert, S., and G. Ondracek. 1988. Sintered Glass. Pp. 161-192 in *Radioactive Waste Forms for the Future*, W. Lutze and R. C. Ewing, eds. Amsterdam: North-Holland.

Garcia, R. S. 1997. Waste Inventories/Characterization Study. (INEL/EXT-97-00600; September). Idaho Falls, Idaho: U.S. Department of Energy/Lockheed Martin Idaho Technologies Company.

Godfrey, R. 1999. Letter (February 16) to Thomas Kiess of the National Research Council, regarding the Australian Nuclear Science and Technology Organization's (ANSTO's) SYNROC ceramics technology. Embassy of Australia Office of the General Counsellor, Washington, D.C.

Gougar, M. L. D., D. D. Siemer, and B. E. Scheetz. 1995. Vitrifiable Concrete for Disposal of Spent Nuclear Fuel Reprocessing Waste at I.N.E.L. Proceedings of the Materials Research Society Annual Meeting Symposium on the Scientific Basis for Nuclear

Waste Management XIX. Idaho Falls, Idaho: Idaho National Engineering and Environmental Laboratory.

Gougar, M. L. D., D. D. Siemer, and B. E. Scheetz. 1996. Vitrifiable concrete for disposal of spent nuclear fuel reprocessing waste at INEL. Pp. 359-366 in Proceedings of the Embedded Topical Meeting on DOE Spent Nuclear Fuel and Fissile Material Management. Reno, NV: American Nuclear Society.

Gougar, M. L. D., B. E. Scheetz, and D. D. Siemer. 1999. A novel waste form for disposal of spent nuclear reprocessing waste: A vitrifiable cement. *Nuclear Technology*, January.

Guerrero A., S. Hérnandez, and S. Goñi. 1998. Durability of cement-based materials in simulated radioactive liquid waste: Effect of phosphate, sulphate and chloride ions. *Journal of Materials Research* 13(8):2151-2160.

Harjula, R., J. Lehto, E. H. Tusa, and A. Paavola. 1994. Industrial Scale Removal of Cesium with Hexacyanoferrate Exchanger-Process Development, *Nuclear Technology* 107:272-284.

Hayward, P. J. 1988. Glass-Ceramics. Pp. 427-493 in *Radioactive Waste Forms for the Future*, W. Lutze and R.C. Ewing, eds. (North-Holland, Amsterdam) 427-493.

Heimann, R. B., and T. T. Vandergraaf. 1988. Cubic zirconia as a candidate waste form for actinades-dissolution studies. *Journal of Nuclear Material* 7:583-586.

Herbst, R. S., D. S. Fryer, K. N. Brewer, C. K. Johnson, and T. A. Todd. 1995. Experimental Results: Pilot Plant Calcine Dissolution and Liquid Feed Stability (INEL/EXT-95/0097; February). Idaho Falls, Idaho: U.S. Department of Energy (DOE)/Lockheed Martin Idaho Technologies Company.

Herbst, R. S., K. N. Brewer, T. G. Garn, J. D. Law, R. T. Tillotson, and T. A. Todd. 1996. A Comparison of TRUEX and CMP Solvent Extraction Processes for Actinide Removal From ICPP Wastes (INEL-96/0094; April). Idaho Falls, Idaho: U.S. Department of Energy/Lockheed Martin Idaho Technologies Company.

Herbst, R. S., J. D. Law, and K. N. Brewer. 1998. Baseline TRUEX Flowsheet Development for the Removal of the Actinides from Dissolved INTEC Calcine Using Centrifugal Contactors (INEL/EXT-98-00833; August). Idaho Falls, Idaho: U.S. Department of Energy/Lockheed Martin Idaho Technologies Company.

Idaho National Engineering and Environmental Laboratory (INEEL). 1997. Annual Update Review Copy INEEL Site Treatment Plan. Idaho Falls, Idaho: U.S. Department of Energy Idaho National Engineering and Environmental Laboratory.

Jostsons, A., E. R. Vance, and R. Hutchings. 1996. Hanford HLW Immobilization in SYNROC (abstract). Sydney, Australia: Australian Nuclear Science and Technology Organization.

Jostsons, A., E. Vance, and G. Durance. 1997. The Role of SYNROC in Partitioning and Conditioning Strategies in Radioactive Waste Management (abstract). Sydney, Australia: Australian Nuclear Science and Technology Organization.

Kimmel, R. 1999a. Letter (January 14) to Tom Kiess of the National Research Council, regarding information request and response to e-mail (EIS-99-002). Idaho Falls, Idaho: U.S. Department of Energy.

Kimmel, R. 1999b. Letter (March 17) to Tom Kiess of the National Research Council, regarding requesting clarification on separations factors (DFs) as described in the Fluor Daniel study (DOE/ID/13206) (EIS-99-027). Idaho Falls, Idaho: U.S. Department of Energy.

Kimmel, R. 1999c. Letter (November 16) to Tom Kiess of the National Research Council, regarding Dr. Tom Kiess' electronic mail, dated November 12, 1999, to A. Olson requesting clarification of earlier submitted data (OPE-EIS-99-077). Idaho Falls, Idaho: U.S. Department of Energy.

Kirk, R., and D. Othmer, eds. 1992. Encyclopedia of Chemical Technology, 4th Ed., 1st Rev. New York: John Wiley & Sons 2:557

Knecht, D. A., P. C. Kong, and T. P. O'Holleran. 1996. Proposed Glass-Ceramic Waste Forms For Immoblizing Excess Plutonium. Department of Energy Spent Nuclear Fuel & Fissile Material Management Meeting in Reno, NV, June 16-20. La Grange Park, Ill.: American Nuclear Society. U.S.

Knecht, D. A., M. D. Staiger, J. D. Christian, C. L. Bendixsen, G. W. Hogg, and J. R. Berreth. 1997. Historical fuel reprocessing and HLW Management in Idaho. *Radwaste Magazine* 4:36-47.

Knecht, D., J. Valentine, A. Luptak, M. Staiger, H. Loo, and T. Wichmann. 1999. Options for Determining Equivalent MTHM for Department of Energy High-Level Waste (INEEL/EXT-99-00317; March). Idaho Falls, Idaho: U.S. Department of Energy Idaho National Engineering and Environmental Laboratory.

Landman, W. H. Jr., and C. M. Barnes. 1998. TRU Separations Options Study Report (INEEL/EXT-97-01428; February). Idaho Falls, Idaho: U.S. Department of Energy/Lockheed Martin Idaho Technologies Company.

Law, J. D., K. N. Brewer, R. S. Herbst, T. A. Todd. 1996. Demonstration of the TRUEX Process for Partitioning of Actinides from Actual ICPP Tank Waste Using Centrifugal Contactors in a Shielded Cell Facility (INEL-96/0353; September). Idaho Falls, Idaho: U.S. Department of Energy/Lockheed Martin Idaho Technologies Company.

Law, J. D., and D. J. Wood. 1996. Development and Testing of a SREX Flowsheet for the Partitioning of Strontium and Lead from Simulated ICPP Sodium-Bearing Waste (INEL-96/0437; November). Idaho Falls, Idaho: U.S. Department of Energy/Lockheed Martin Idaho Technologies Company.

Law, J. D., D. J. Wood, L. G. Olson, and T. A. Todd. 1997. Demonstration of a SREX Flowsheet for the Partitioning of Strontium and Lead from Actual ICPP Sodium-Bearing Waste (INEEL/EXT-97-00832; August). Idaho Falls, Idaho: U.S. Department of Energy /Lockheed Martin Idaho Technologies Company.

Law, J. D., K. N. Brewer, R. S. Herbst, T. A. Todd, and L. G. Olson. 1998. Demonstration of an Optimized TRUEX Flowsheet for Partitioning of Actnides from Actual ICPP Sodium-Bearing Waste Using Centrifugal Contactors in a Shielded Cell Facility (INEEL/EXT-98-00004, January). Idaho Falls, Idaho: U.S. Department of Energy /Lockheed Martin Idaho Technologies Company.

Lee, A. E., and D. D. Taylor. 1998. Cementitious Waste Option Scoping Study Report (INEEL/EXT-97-01400; February). Idaho Falls, Idaho: U.S. Department of Energy /Lockheed Martin Idaho Technologies Company.

Lockheed Martin Idaho Technologies Company (LMITCO). 1996. High-Level Waste Program Plan (INEL-96/122; April). Idaho Falls, Idaho: U.S. Department of Energy/LMITCO.

Lodge, E. J. 1995. Settlement Agreement between *Public Service Co. of Colorado v. Batt*, No.CV 91-00350S0EJL (D. Id.) and *United States v. Batt*, No. CV-91-0054-S-EJL (D. Id.) Consent Order signed October 17, 1995, by Edward J. Lodge.

Lopez, D. A., and R. R. Kimmitt. 1998. Vitrified Waste Option Study Report (INEEL/EXT-97-01389; February). Idaho Falls, Idaho: U.S. Department of Energy/Lockheed Martin Idaho Technologies Company.

Lutze, W., and R. C. Ewing, eds. 1988. *Radioactive Waste Forms for the Future*. Amsterdam: North-Holland.

Lutze, W. 1998. Candidate Waste Form(s) for INEEL Calcines. Presentation given at the National Research Council Committee's INEEL High-Level Waste Alternative

Treatments Review meeting on October 1, at the Idaho Nuclear Technology and Engineering Center (INTEC). Idaho Falls, Idaho.

Mann, N. R., and T. A. Todd. 1998. Evaluation and Testing of the Cells Unit Crossflow Filter on INEEL Dissolved Calcine Slurries (INEEL/EXT-98-00749; August). Idaho Falls, Idaho: U.S. Department of Energy/Lockheed Martin Idaho Technologies Company.

McDaniel, E. W., and D. B. Delzer. 1988. FUETAP Concrete. Pp. 565-588 in *Radioactive Waste Forms for the Future*, W. Lutze and R. C. Ewing, eds. Amsterdam: North-Holland.

McNeese, L. E., J. B. Berry, G. E. Butterworth III, E. D. Collins, and T. H. Monk. 1988. Overall strategy and program plan for management of radioactively contaminated liquid wastes and transuranic sludges at the Oak Ridge National Laboratory (ORNL) (ORNL/TM-10757; December). Oak Ridge, TN: ORNL.

Miller, C. J., A. L. Olson, and C. K. Johnson. 1997. Cesium absorption from acidic solutions using ammonium molybdophosphate on a polyacrylonitrile support (AMP-PAN). *Separation Science and Technology*. 32:1-4.

Moore, J. G., H. W., Godbee, A. H. Kibbey, and D. S. Joy. 1975. Development of Cementitious Grouts for the Incorporation of Radioactive Wastes, Part I: Leach Studies (ORNL-4962). Oak Ridge, Tenn.: Oak Ridge National Laboratory.

Moore, J. G. 1981. A survey of concrete waste forms. Proceedings of a Conference on Alternative Nuclear Waste Forms and Interactions in Geologic Media (NTIS 194-216), J. G. Moore, L. A. Boatner, and G. C. Battle, eds. Springfield, VA: National Technical Information Service.

Murphy, J. A., L. Pincock, and I. N. Christiansen. 1995. ICPP Radioactive Liquid and Calcine Waste Technologies Evaluation Final Report and Recommendation (INEL-94/0119; April). Idaho Falls, Idaho: U.S. Department of Energy/Idaho National Engineering and Environmental Laboratory.

National Research Council (NRC). 1965. The Radiochemistry of Plutonium. NAS-NS 3058. G. H. Coleman, ed. Clearinghouse for Federal Scientific and Technical Information, National Bureau of Standards, U.S. Department of Commerce, Springfield, Virginia.

NRC. 1999. An End State Methodology for Identifying Technology Needs for Environmental Management, with an Example from the Hanford Site Tanks. Washington, D.C.: National Academy Press.

Oak Ridge National Laboratory (ORNL). 1996. Integrated Data Base Report—1995: U.S. Spent Nuclear Fuel and Radioactive Waste Inventories, Projections, and Characteristics. (DOE/RW-0006, Rev. 12 – December). Washington, D.C.: U.S. Department of Energy.

Office of Civilian Radioactive Waste Management. 1998. Civilian Radioactive Waste Management System Requirements Document Revision 4 (DOE/RW-0406, Rev4, May). Washington, D.C.: U.S. Department of Energy Office of Civilian Radioactive Waste Management.

Olson, A. L. 1997. Planned Cs Removal Technology (ALO-02-97; interdepartmental communication). Idaho Falls, Idaho: U.S. Department of Energy/Lockheed Martin Idaho Technologies Company.

Olson, A. L. 1998. Endstate Waste Form Requirements and Fates of Important Waste Constituents. Presentation given at the Idaho National Engineering and Environmental Laboratory (INEEL) Committee meeting on October 1-2, in Idaho Falls, Idaho.

Organisation for Economic Co-operation and Development (OECD) Nuclear Energy Agency (NEA). 1995. The Environmental and Ethical Basis of Geological Disposal of Long-Lived Radioactive Wastes. Paris, France: OECD NEA.

Oversby, V. M., C. C. McPheeters, C. Degueldre, and J. M. Paratte. 1997. Control of civilian plutonium inventories using burning in a non-fertile fuel. *Journal of Nuclear Materials* 245:12-26.

Palmer, W. B., M. J. Beer, M. Cukars, J. P. Law, C. B. Millet, J. A. Murphy, J. A. Nenni, C. V. Park, J. I. Pruitt, E. C. Thiel, F. S. Ward, and J. Woodard. 1994. ICPP Tank Farm Systems Analysis (WINCO-1192; January). Washington, DC: Westinghouse Idaho Nuclear Company, Inc.

Palmer, W. B. 1996. HLW Treatment Alternatives Evaluation (WBP-29-96 Draft Report; August 16). Idaho Falls, Idaho: U.S. Department of Energy/Lockheed Martin Idaho Technologies Company.

Palmer, B. 1998. HLW Baseline Plan, Accelerating Cleanup: Paths to Closure. Presentation given at the National Research Council Committee's INEEL High-Level Waste Alternative Treatments Review meeting on August 17, at the Idaho Nuclear Technology and Engineering Center (INTEC). Idaho Falls, Idaho.

Palmer, B., and A. L. Olson. 1998. Sodium Bearing Waste Treatment Study. Presentation given at the NRC Committee's INEEL High-Level Waste Alternative Treatments Review meeting on August 17, at INTEC. Idaho Falls, Idaho.

Palmer, W. B., C. B. Millet, M. D. Staiger, and F. S. Ward. 1998. ICPP Tank Farm Planning through 2012 (INEEL/EXT-98-00339; April). Idaho Falls, Idaho: U.S. Department of Energy/Lockheed Martin Idaho Technologies Company.

Parker, S. P. ed. 1994. Dictionary of Scientific and Technical Terms, Fifth Edition. New York McGraw-Hill.

Post, R. G. 1981. Independent Review Panel to Evaluate and Review Alternative Waste Forms to Immobilize the Idaho Chemical Processing Plant (ICPP) Calcined Waste, Final Report (ENICO-1088; December 1980). Idaho Falls, Idaho: U.S. Department of Energy Idaho Operations Office & ENICO.

Public Law (P. L.) 97-425 (1982) as amended by P. L. 100-203 (1987) and P. L. 102-486 (the Energy Policy Act of 1992).

Raytheon Engineers & Constructors. 1994. Idaho Chemical Processing Plant Waste Immobilization Facility Feasibility Study Report, October 1994, Volume I—Feasibility Design Description (MS511.163; October). Idaho Falls, Idaho: U.S. Department of Energy Idaho Operations Office.

Rechard, R., ed. 1995. Performance Assessment of the Direct Disposal in Unsaturated Tuff of Spent Nuclear Fuel and High-Level Waste Owned by U.S. Department of Energy, Volumes 1-3. Albuquerque, NM: Sandia National Laboratories.

Russell, N. E., and A. D. Taylor. 1998. Hot Isostatic Press Waste Option Study Report (INEEL/EXT-97-01392; February). Idaho Falls, Idaho: U.S. Department of Energy/Lockheed Martin Idaho Technologies Company.

Russell, N. E., T. G. McDonald, J. Banaee, C. M. Barnes, L. W. Fish, S. J. Losinski, H. K. Peterson, J. W. Sterbentz and D. R. Wenzel. 1998. Waste Disposal Options Report (INEEL/EXT-97-01145; February). Idaho Falls, Idaho: U.S. Department of Energy/Lockheed Martin Idaho Technologies Company.

Sales, B. C., and L. A. Boatner. 1988. Lead-iron phosphate glass. Pp. 193-232 in *Radioactive Waste Forms for the Future*, W. Lutze and R. Ewing, eds. Amsterdam: North-Holland.

Schindler, R. E. 1974. Revised Design Criteria for ICPP Fourth Calcined Solids Storage Facility (September). Idaho Falls, Idaho: Allied Chemical Corporation, Idaho Chemical Programs, Operations Office.

Siemer, D. D., M. W. Grutzeck, and B. E. Scheetz. In Press. Paper on Comparison of Materials for Making Hydroceramic Waste Forms (abstract).

Siemer, D. D. 1995. Hot Isostatically Pressed Concrete as a Radwaste Form (abstract). *Environmental Issues and Waste Management Technologies* 61:657-664.

Siemer, D. D. 1995. Hot Isostatic Press (HIP) Vitrification of Radwaste Concretes. Pp. 403-410 in Materials Research Society Symposium Proceedings: Scientific Basis for Nuclear Waste Management XIX, Vol. 412. Boston, Mass.: Materials Research Society.

Siemer, D. D., D. M. Roy, M. W. Grutzeck, M. L. D. Gougar, and B. E. Scheetz. 1997. PCT Leach Tests of Hot Isostatically Pressed (HIPped) Zeolitic Concretes. Pp. 303-310 in Materials Research Society Symposium Proceedings: Scientific Basis for Nuclear Waste Management XX., Vol. 465. Boston, Mass.: Materials Research Society.

Siemer, D. D., M. W. Grutzeck, D. M. Roy, and B. E. Scheetz. 1998. Zeolite Waste Forms Synthesized From Sodium Bearing Waste and Metakaolinite. Waste Management '98. Session 36, Paper 2. Available at URL: http://www.wmsym.org/wm98/html/sess36/36-02/36-02.htm.

Slaughterbeck, D. C., W. E. House, G. A. Freund, T. D. Enyeart, E. C. Benson, Jr., and K. D. Bulhman. 1995. Accident Assessments for Idaho National Engineering Laboratory Facilities. Science Applications International Corporation DOE/ID-10471. Idaho Falls, Idaho: U.S. Department of Energy.

Sombret, C. 1998a. Letter (April 25) from C. Sombret to Tom Kiess of the National Research Council regarding some remarks about the fate of the INEEL calcined HLW. France.

Sombret, C. 1998b. E-mail (July 2) from C. Sombret to Tom Kiess of the National Research Council, regarding INEEL calcine. France.

Spaulding, B. C., R. A. Gavalya, M. M. Dahlmeir, L. C. Tuott, K. D. McAllister, K. C. DeCoria, S. P. Swanson, R. D. Adams, G. C. McCoy, R. J. Turk. 1998. ICCP Tank Farm Closure Study, Volumes I-III. Idaho Falls, Idaho: U.S. Department of Energy/Lockheed Martin Idaho Technologies Company.

Staples, B. A., G. S. Pomiak, and E. L. Wade. 1979. Properties of Radioactive Calcine Retrieved from the Second Calcined Solids Storage Facility at ICPP: Idaho Chemical Programs (UC-70—ICP-1189; March). Idaho Falls, Idaho: U.S. Department of Energy/Allied Chemical Corporation.

Staples B., H. Cole, and D. Pavlica. 1983. Properties of formula 127 glass prepared with radioactive zirconia calcine, p.125 in Scientific Basis for Nuclear Waste Management VI, ed. D. G. Brookins, Materials Research Society Symposium Proceedings V.15. Elsevier Science Publishing Co., Inc.

Staples, B. A., T. P. O'Holleran, D. A. Knecht, P. C. Kong, S. J. Johnson, K. Vinjamuri, B. A. Scholes, A. W. Erickson, S. M. Frank, M. K. Meyer, M. Hansen, E. L. Wood, and H. C. Wood. 1997. Preparation of Plutonium Waste Forms with ICPP Calcined High-Level Waste (INEEL/EXT-97-00496). Idaho Falls, Idaho: U.S. Department of Energy/Lockheed Martin Idaho Technologies Company.

Staples, B. 1998. Twenty Years of INTEC Waste Form Development. Presentation given at the NRC Committee's INEEL High-Level Waste Alternative Treatments Review meeting on August 18, at the Idaho Nuclear Technology and Engineering Center (INTEC). Idaho Falls, Idaho.

Staples, B. A., D. K. Peeler, G. F. Piepel, J. D. Vienna, B. A. Scholes, and C. A. Musick. 1998. The Preparation and Characterization of INTEC HAW Phase 1 Composition Variation Study Glasses (INEEL/EXT-98-00970; September). Idaho Falls, Idaho: U.S. Department of Energy/Lockheed Martin Idaho Technologies Company.

TFA (Tanks Focus Area). 1999. TFA Technical Highlights: Period Ending January 31, 1999. U.S. Department of Energy/Pacific Northwest National Laboratory.

Thompson, M. In Press. Pretreatment Radionuclide Separations of Cs/Tc from Supernates (Document No. WSR-MS-98-00601). Savannah, Ga.: U.S. Department of Energy/Westinghouse.

Todd, T. A., K. N. Brewer, D. J. Wood, P. A. Tullock, L. G. Olson, and A. L. Olson. In Press. Evaluation and testing of inorganic ion exchange sorbents for the removal of Cesium-137 from Idaho Chemical Processing Plant Acidic Tank Waste. (abstract). *Separation Science & Technology*.

Ullman, F., ed. 1985. *Encyclopedia of Industrial Chemistry*. Fifth Edition, 1 Rev. Vol. A1. VCH.

U.S. Department of Energy (U.S. DOE). 1985. An Evaluation of Commercial Repository Capacity for the Disposal of Defense High-Level Waste. Washington, D.C.: U.S. DOE.

U.S. DOE. 1987. Proposed Method for Assigning Metric Tons of Heavy Metal Values to Defense High-Level Waste Forms to be Disposed of in a Geologic Repository (DOE/RL-87-04 – August). Richland, WA: U.S. DOE.

U.S. DOE. 1996a. Waste Acceptance Criteria for the Waste Isolation Pilot Plant. Westinghouse Electric Corporation Nuclear Fuel Business Unit Waste Isolation Division. (DOE/WIPP-069 Revision 5; April). Washington, D.C.: DOE.

U.S. DOE. 1996c. Waste Acceptance Product Specifications (WAPS) For Vitrified High-Level Waste Forms. EM-WAPS Rev. 02. U.S. DOE Office of Environmental Management.

U.S. DOE. 1997. Nevada Test Site Waste Acceptance Criteria (DOE/NV-335 NTSWAC, Revision 1; August). Reno, NV: U.S. DOE Nevada Operations Office Waste Management Division.

U.S. DOE. 1998c. DOE Awards Contract. *Environmental Update* Summer (18).

U.S. DOE Alternatives Team. 1998. Process for Identifying Potential Alternatives for the INEEL High-Level Waste and Facilities Disposition Environmental Impact Statement (Predecisional Draft) DOE-ID 10627. Idaho Falls, Idaho: U.S. DOE.

Vinjamuri, K. 1995a. Glass Waste Forms for the Na-bearing High Activity Waste Fractions. (INEL-95/0214; June). Idaho Falls, Idaho: U.S. Department of Energy/Lockheed Martin Idaho Technologies Company.

Wichmann, T., N. Brooks, and M. Heiser. 1996. Regulatory Analysis and Proposed Path Forward for the Idaho National Engineering Laboratory High-Level Waste Program. (DOE/ID-10544, Revision 1; October). Idaho Falls, Idaho: U.S. Department of Energy/Lockheed Martin Idaho Technologies Company.

Wichmann, T. 1998. High-Level Waste Regulatory Framework and Issues. Presentation given at the NRC Committee's INEEL High-Level Waste Alternative Treatments Review meeting on August 18, at the Idaho Nuclear Technology and Engineering Center (INTEC). Idaho Falls, Idaho.

Wichmann, T. 1998. NAS Regulatory Framework. Presentation given at the NRC Committee's INEEL High-Level Waste Alternative Treatments Review meeting on October 1-2, at the Idaho U.S. Department of Energy Operations Office. Idaho Falls, Idaho.

Wood, D. J., J. D. Law, T. G. Garn, R. D. Tillotson, P. A. Tullock, and T. A. Todd. 1997. Development of the SREX Process for the Treatment of ICPP Liquid Wastes (INEEL/EXT-97/00831; October). Idaho Falls, Idaho: U.S. Department of Energy/Lockheed Martin Idaho Technologies Company.

Appendix A

Documents Received During This Study

Alternative Waste Form Peer Review Panel. 1979. The Evaluation and Review of Alternative Waste Forms for Immobilization of High Level Radioactive Wastes. Washington, D.C.: U.S. Department of Energy Office of Nuclear Waste Management.

Australian Nuclear Science and Technology Organization (ANSTO). 1997–1998. ANSTO's Annual Report. Sydney, Australia.

Barnes, C. M., D. D. Taylor, and B. R. Helm. 1997. Transmittal of Process Basis Information for the Feasibility Study of the Preferred Alternative for Treatment of ICCP Sodium-Bearing Wastes, Calcine and Low-Level Wastes (CMB-01-97; interdepartmental communication, April). Idaho Falls, Idaho: U.S. Department of Energy/Lockheed Martin Idaho Technologies Company.

Barrer, R. M., and J. F. Cole. 1970. Chemistry of soil minerals. Part VI. Salt entrainment by sodalite and cancrinite during their synthesis. *Chemical Society Journal*. No.s 2:7–12 Pp. 1516-1523.

Barrer, R. M., J. F. Cole, and H. Villiger. 1970. Chemistry of soil minerals. Part VII. Synthesis, properties, and crystal structures of salt-filled cancrinites. *Chemical Society Journal* No.s 2:7–12. Pp. 1523-1531.

Bell, J. T. 1998. Alternatives to HLW Vitrification: The Need for Common Sense. Presentation given at SPECTRUM '98 September 13-18, in Denver, Colorado. Bell Consultants.

Berreth, J. R. 1988. Inventories and Properties of ICPP Calcined High-Level Waste. (WINCO 1050; February). Idaho Falls, Idaho: U.S. Department of Energy/Westinghouse Idaho Nuclear Company, Inc.

Bonano, E. J., M. S. Y. Chu, L. L. Price, S. H. Conrad, and P. T. Dickman. 1991. The Disposal of Orphan Wastes Using the Greater Confinement Disposal Concept (DE-Ac04-76DP00789). Sandia National Laboratories Albuquerque, NM: U.S. Department of Energy.

Bonniaud, R., P. Labe, and C. Sombret. 1968. Vitrification des residus solides fluores provenant de l'attaque par le fluor des combustibles irradies. Presentation given at the Symposium sur le Retraitement par Voie Seche on October 29 in Mol, Belgium.

Bosley, J. 1998. Closure Study for the Liquid Effluent Treatment and Disposal Facility Building CPP-1618 (INEEL/EXT-98-01152; February). Idaho Falls, Idaho: U.S. Department of Energy/Lockheed Martin Idaho Technologies Company.

Brass, L. 1998. Foster Wheeler discusses project. Oak Ridge, Tenn.: *The Oak Ridger*. October 15.

Brewer, K. N., R. S. Herbst, T. J. Tranter, A. L. Olson, T. A. Todd, I. D. Goodwin, and C. W. Lundholm. 1995. Dissolution of two NWCF calcines: extent of dissolution and characterization of undissolved solids. Idaho National Engineering Laboratory (INEL-95/0098; February). Idaho Falls, Idaho: U.S. Department of Energy/Lockheed Martin Idaho Technologies Company.

Brewer, K. N., T. A. Todd, A. L. Olson, D. J. Wood, V. M. Gelis, E. A. Kozlitin, and V. V. Milyutin. 1996. Cesium removal from radioactive idaho chemical processing plant acidic waste with potassium copper hexacyanoferrate (INEL-96/0356; November). Idaho Falls, Idaho: U.S. Department of Energy/Lockheed Martin Idaho Technologies Company.

Brewer, K. N., R. D. Tillotson, P. A. Tullock, T. G. Garn, R. S. Herbst, J. D. Law, T. A. Todd, and A. L. Olson. 1997. Elimination of Phosphate and Zirconium in the High-Activity Fraction Resulting from TRUEX Partitioning of ICPP Zirconium Calcines. Idaho National Engineering Laboratory (INEEL/EXT-97-00836; July). Idaho Falls, Idaho: U.S. Department of Energy/Lockheed Martin Idaho Technologies Company.

Brewer, K. N., A. L. Olson, W. S. Roesener, and J. L. Tonso. 1997. Experimental results of calcine dissolution studies performed during FY-94, 95 (INEL-97-01192; September). Idaho Falls, Idaho: U.S. Department of Energy/Lockheed Martin Idaho Technologies Company.

Brotzman J. R. 1978. Vitrification of high-level alumina nuclear waste. P. 215 in *Scientific Basis for Nuclear Waste Management II*, C. J. M. Northrup, Jr., ed. New York: Plenum Press.

Brownell, L. E., C. H. Kindle, and T. L. Theis. 1973. Review of Literature Pertinent to the Aqueous Conversion of Radionuclides to Insoluble Silicates with Selected References and Bibliography (Revised) (ARH-2731 REV; December). Richland, Wash.: U.S. Atomic Energy Commission.

Campbell, D. O., and D. D. Lee. 1991. Treatment Options and Flowsheets for ORNL Low-Level Liquid Waste Supernate (December - ORNL/TM-11800). Oak Ridge, Tenn.: Oak Ridge National Laboratory.

Campbell, D. O., D. D. Lee, and T. A. Dillow. 1991. Development Studies for the Treatment of ORNL Low-Level Liquid Waste (November - ORNL/TM-11798). Oak Ridge, Tenn.: Oak Ridge National Laboratory.

Chen, F., and D. E. Day. 1999. Corrosion of Selected Refractories by Iron Phosphate Melts, *Ceramic Transactions* 93:213-218.

Childs, K. F., R. I. Donovan, and M. C. Swenson. 1982. The ninth processing campaign in the waste calcining facility. (DE-AC07-79IDO1675; April). Idaho Falls, Idaho: U.S. Department of Energy.

Cochran, J. R. 1997. Letter (March 20) to Darryl Siemer of LMITCO, regarding Cementitious High-Level Waste and 40 CFR 268.40. Albuquerque, N. Mex.: Sandia National Laboratories. U.S. Department of Energy.

Cochran, J., B. Crowe, and F. DiSanza. 1999. Intermediate depth burial of classified transuranic wastes in arid alluvium (abstract). Albuquerque, N.Mex.: Sandia National Laboratories.

Cole, H. S., and E. L. Colton. 1982. Comparison of candidate glasses to vitrify ICPP calcined alumina waste. *Transactions of the American Nuclear Society* 43:109-110.

Dafoe, R. E., and S. J. Losinski. 1998. Direct Cementitious Waste Option Study Report (INEEL/EXT-97-01399; February). Idaho Falls, Idaho: U.S. Department of Energy/Lockheed Martin Idaho Technologies Company.

Dafoe, R. E., D. A. Lopez, and K. L. Williams. 1997. DPC Loading Feasibility Study Report (INEEL/EXT-97-01251; November). Idaho Falls, Idaho: U.S. DOE/Lockheed Martin Idaho Technologies Company.

Dahlmeir, M. M., L. C. Tuott, B. C. Spaulding, S. P. Swanson, K. D. McAllister, K. C. Decoria, S. L. Coward, R. J. Turk, R. A. Adams, and C. W. Barnes. 1998. Calcined Solids Storage Facility Closure Study (INEEL/EXT-97-01396 February). Idaho Falls, Idaho: U.S. Department of Energy/Lockheed Martin Idaho Technologies Company.

Davis, P. 1993. Letter (March 2) to Darryl Siemer of Westinghouse Idaho Nuclear Company, regarding greater confinement disposal of radioactive waste. Albuquerque, NM: Sandia National Laboratories. U.S. Department of Energy.

Day, D. E., Z. Wu, C. S. Ray, and P. Hrma. 1999. Chemically durable iron phosphate glass wasteforms. *Journal of Non-Crystalline Solids* 241:1-12.

Delegard, C. H., and G. S. Barney. 1975. Fixation of Radioactive Waste by Reaction With Clays: Progress Report (ARH-2731 REV; July). Richland, Wash.: U.S. Energy Research and Development Administration/Atlantic Richfield Hanford Company.

Dirk, W. J. 1994. Long-Term Laboratory Corrosion Monitoring of Calcine Bin Set Materials Exposed to Zirconia Calcine. Idaho Falls, Idaho: U.S. Department of Energy/Westinghouse Idaho Nuclear Company, Inc.

Dole, L. R., G. C. Rogers, M. T. Morgan, D. P. Stinton, J. H. Kessler, S. M. Robinson, and J. G. Moore. 1983. Cement-Based Radioactive Waste Hosts Formed under Elevated Temperatures and Pressures (Fuetap Concretes) for Savannah River Plant High-Level Defense Waste (ORNL/TM-8579; March). Oak Ridge, TN: Oak Ridge National Laboratory.

Fluor Daniel, Inc. 1996. Sodium Waste Alternative Treatment and Disposal Feasibility Study (94-29-FT-008; April 1996 and addendum of July 1996). Idaho Falls, Idaho: U.S. Department of Energy Idaho Field Office.

Fluor Daniel, Inc. 1997. Idaho Chemical Processing Plant Waste Treatment Facilities Feasibility Study Report, Volumes I & II. Idaho Falls, Idaho: DOE-Idaho Field Office.

Fluor Daniel Hanford, Inc. 1998. Hanford Site Solid Waste Acceptance Criteria for June 29 (Document No. HNF-EP-0063). Available at URL: http//www.hanford.gov.

Gahlert, S., and G. Ondracek. 1988. Sintered Glass. Pp. 161-192 in *Radioactive Waste Forms for the Future*, W. Lutze and R. C. Ewing, eds. Amsterdam: North-Holland.

Garcia, R. S. 1997. Waste Inventories/Characterization Study. (INEL/EXT-97-00600; September). Idaho Falls, Idaho: U.S. DOE/Lockheed Martin Idaho Technologies Company.

Godfrey, R. 1999. Letter (February 16) to Thomas Kiess of the National Research Council, regarding the Australian Nuclear Science and Technology Organization's (ANSTO's) SYNROC ceramics technology. Embassy of Australia Office of the General Counsellor, Washington, D.C.

Gougar, M. L. D., D. D. Siemer, and B. E. Scheetz. 1995. Vitrifiable Concrete for Disposal of Spent Nuclear Fuel Reprocessing Waste at I.N.E.L. Proceedings of the Materials Research Society Annual Meeting Symposium on the Scientific Basis for Nuclear Waste Management XIX. Idaho Falls, Idaho: Idaho National Engineering and Environmental Laboratory.

Gougar, M. L. D., D. D. Siemer, and B. E. Scheetz. 1996. Vitrifiable concrete for disposal of spent nuclear fuel reprocessing waste at INEL. Pp. 359-366 in Proceedings of the Embedded Topical Meeting on DOE Spent Nuclear Fuel and Fissile Material Management. Reno, NV: American Nuclear Society.

Gougar, M. L. D., B. E. Scheetz, and D. D. Siemer. 1999. A novel waste form for disposal of spent nuclear reprocessing waste: A vitrifiable cement. *Nuclear Technology*, January.

Guerrero A., S. Hérnandez, and S. Goñi. 1998. Durability of cement-based materials in simulated radioactive liquid waste: Effect of phosphate, sulphate and chloride ions. *Journal of Materials Research* 13(8):2151-2160.

Grutzeck, M. W., and D. D. Siemer. 1997. Zeolites synthesized from class F fly ash and sodium aluminate slurry. *Journal of the American Ceramic Society* 809:2449-2453.

Haley, D. J. 1998. Engineering Study Report for RCRA Facility Closure of CPP-604 (Process Equipment Waste Evaporator) and CPP-605 (INEEL/EXT-98-00151; February). Idaho Falls, Idaho: U.S. Department of Energy/Lockheed Martin Idaho Technologies Company.

Harjula, R., J. Lehto, E. H. Tusa, and A. Paavola. 1994. Industrial Scale Removal of Cesium with Hexacyanoferrate Exchanger-Process Development, *Nuclear Technology* (September) 107:272-284.

Hayward, P. J. 1988. Glass-Ceramics Pp. 427-493 in *Radioactive Waste Forms for the Future*, W. Lutze and R. C. Ewing, eds. Amsterdam: North-Holland.

Heimann, R. B., and T. T. Vandergraaf. 1988. Cubic zirconia as a candidate waste form for actinades-dissolution studies. *Journal of Nuclear Material* 7:583-586.

Heiser, M. B. 1998. Tank Closure. Presentation given at the NRC Committee's INEEL High-Level Waste Alternative Treatments Review meeting on October 1-2, at the U.S. DOE Idaho Operations Office, Idaho Falls, Idaho.

Herbst, R. S., D. S. Fryer, K. N. Brewer, C. K. Johnson, and T. A. Todd. 1995. Experimental Results: Pilot Plant Calcine Dissolution and Liquid Feed Stability (INEL/EXT-95/0097; February). Idaho Falls, Idaho: U.S. Department of Energy (DOE)/Lockheed Martin Idaho Technologies Company.

Herbst, R .S., K. N. Brewer, T. G. Garn, J. D. Law, R. T. Tillotson, and T. A. Todd. 1996. A Comparison of TRUEX and CMP Solvent Extraction Processes for Actinide Removal From ICPP Wastes (INEL-96/0094; April). Idaho Falls, Idaho: U.S. DOE/Lockheed Martin Idaho Technologies Company.

Herbst, R. S., J. D. Law, K. N. Brewer, T. A. Todd, and L. G. Olson. 1997. TRUEX Flowsheet Testing for the Removal of the Actinides from Dissolved ICPP Zirconium Calcine Using Centrifugal Contactors (INEL/EXT-97-00837; December). Idaho Falls, Idaho: U.S. DOE/Lockheed Martin Idaho Technologies Company.

Herbst, R. S., J. D. Law, and K. N. Brewer. 1998. Baseline TRUEX Flowsheet Development for the Removal of the Actinides from Dissolved INTEC Calcine Using Centrifugal Contactors (INEL/EXT-98-00833; August). Idaho Falls, Idaho: U.S. DOE/Lockheed Martin Idaho Technologies Company.

Idaho National Engineering and Environmental Laboratory (INEEL). 1997. Annual Update Review Copy INEEL Site Treatment Plan. Idaho Falls, Idaho: U.S. Department of Energy Idaho National Engineering and Environmental Laboratory.

Jostsons, A., E. R. Vance, and R. Hutchings. 1996. Hanford HLW Immobilization in SYNROC (abstract). Sydney, Australia: Australian Nuclear Science and Technology Organization.

Jostsons, A., E. Vance, and G. Durance. 1997. The Role of SYNROC in Partitioning and Conditioning Strategies in Radioactive Waste Management (abstract). Sydney, Australia: Australian Nuclear Science and Technology Organization.

Kimmel, R. 1999a. Letter (January 14) to Tom Kiess of the National Research Council, regarding information request and response to e-mail (EIS-99-002). Idaho Falls, Idaho: U.S. Department of Energy.

Kimmel, R. 1999b. Letter (March 17) to Tom Kiess of the National Research Council, regarding requesting clarification on separations factors (DFs) as described in the Fluor Daniel study (DOE/ID/13206) (EIS-99-027). Idaho Falls, Idaho: U.S. Department of Energy.

Kimmel, R. 1999c. Letter (November 16) to Tom Kiess of the National Research Council, regarding Dr. Tom Kiess' electronic mail, dated November 12, 1999, to A. Olson requesting clarification of earlier submitted data (OPE-EIS-99-077). Idaho Falls, Idaho: U.S. Department of Energy.

Kirk, R., and D. Othmer, eds. 1992. Encyclopedia of Chemical Technology, 4th Ed., 1st Rev. New York: John Wiley & Sons 2:557

Knecht, D. A., P. C. Kong, and T. P. O'Holleran. 1996. Proposed Glass-Ceramic Waste Forms For Immoblizing Excess Plutonium. DOE Spent Nuclear Fuel & Fissile Material Management Meeting in Reno, NV, June 16-20. La Grange Park, Ill.: American Nuclear Society.

Knecht, D. A., M. D. Staiger, J. D. Christian, C. L. Bendixsen, G. W. Hogg, and J. R. Berreth. 1997. Historical fuel reprocessing and HLW Management in Idaho. *Radwaste Magazine* 4:36-47.

Knecht, D. A., K. Vinjamuri, S. V. Raman, B. A. Staples, and J. D. Grandy, S. Johnson, T. P. O'Holleran, and S. Frank. 1998. Summary of HLW Glass-Ceramic Waste Forms Developed in Idaho for Immobilizing Plutonium (abstract). Presentation given at the Waste Management '98 conference on March 2 in Tucson, Arizona.

Knecht, D. 1998. Presentation given at National Research Council Committee's INEEL High-Level Waste Alternative Treatments Review meeting on August 17, at the Idaho Nuclear Technology and Engineering Center (INTEC). Idaho Falls, Idaho.

Knecht, D., J. Valentine, A. Luptak, M. Staiger, H. Loo, and T. Wichmann. 1999. Options for Determining Equivalent MTHM for DOE High-Level Waste (INEEL/EXT-99-00317; March). Idaho Falls, Idaho: U.S. Department of Energy Idaho National Engineering and Environmental Laboratory.

Landman, Jr., W. H., P. W. Bragassa, M. Spinti, M. Pope, and K. Wells. 1997. New Waste Calcining Facility Deactivation Option for Low-Level Waste Grout Disposal (INEEL/EXT-97-01076; December). Idaho Falls, Idaho: U.S. Department of Energy/Lockheed Martin Idaho Technologies Company.

Landman, W. H. Jr., and C. M. Barnes. 1998. TRU Separations Options Study Report (INEEL/EXT-97-01428; February). Idaho Falls, Idaho: U.S. Department of Energy/Lockheed Martin Idaho Technologies Company.

Law, J. D., K. N. Brewer, R. S. Herbst, T. A. Todd. 1996. Demonstration of the TRUEX Process for Partitioning of Actinides from Actual ICPP Tank Waste Using Centrifugal Contactors in a Shielded Cell Facility (INEL-96/0353; September). Idaho Falls, Idaho: U.S. Department of Energy/Lockheed Martin Idaho Technologies Company.

Law, J. D., and D. J. Wood. 1996. Development and Testing of a SREX Flowsheet for the Partitioning of Strontium and Lead from Simulated ICPP Sodium-Bearing Waste (INEL-96/0437; November). Idaho Falls, Idaho: U.S. Department of Energy/Lockheed Martin Idaho Technologies Company.

Law, J. D., D. J. Wood, L. G. Olson, and T. A. Todd. 1997. Demonstration of a SREX Flowsheet for the Partitioning of Strontium and Lead from Actual ICPP Sodium-Bearing Waste (INEEL/EXT-97-00832; August). Idaho Falls, Idaho: U.S. Department of Energy /Lockheed Martin Idaho Technologies Company.

Law, J. D., K. N. Brewer, R. S. Herbst, T. A. Todd, and L. G. Olson. 1998. Demonstration of an Optimized TRUEX Flowsheet for Partitioning of Actinides from Actual ICPP Sodium-Bearing Waste Using Centrifugal Contactors in a Shielded Cell Facility (INEEL/EXT-98-00004, January). Idaho Falls, Idaho: U.S. Department of Energy /Lockheed Martin Idaho Technologies Company.

Lee, A. E., and D. D. Taylor. 1998. Cementitious Waste Option Scoping Study Report (INEEL/EXT-97-01400; February). Idaho Falls, Idaho: U.S. Department of Energy /Lockheed Martin Idaho Technologies Company.

Lockheed Martin Idaho Technologies Company (LMITCO). 1996. High-Level Waste Program Plan (INEL-96/122; April). Idaho Falls, Idaho: U.S. Department of Energy/LMITCO.

Lockheed Martin Idaho Technologies Company. 1998. Accelerating Cleanup: Paths to Closure Idaho Operations Office (DOE/EM0342, PLN-177; February) Idaho Falls, Idaho: U.S. Department of Energy Idaho Operations Office.

Lodge, E. J. 1995. Settlement Agreement between *Public Service Co. of Colorado v. Batt*, No.CV 91-00350S0EJL (D. Id.) and *United States v. Batt*, No. CV-91-0054-S-EJL (D. Id.). Consent Order signed October 17, 1995, by Edward J. Lodge.

Lopez, D. A., and R. R. Kimmitt. 1998. Vitrified Waste Option Study Report (INEEL/EXT-97-01389; February). Idaho Falls, Idaho: U.S. Department of Energy/Lockheed Martin Idaho Technologies Company.

Lutze, W., and R. C. Ewing, eds. 1988. *Radioactive Waste Forms for the Future*. Amsterdam: North-Holland.

Lutze, W. 1998. Candidate Waste Form(s) for INEEL Calcines. Presentation given at the National Research Council Committee's INEEL High-Level Waste Alternative Treatments Review meeting on October 1, at the Idaho Nuclear Technology and Engineering Center (INTEC). Idaho Falls, Idaho.

Mann, N. R., and T. A. Todd. 1998. Evaluation and Testing of the Cells Unit Crossflow Filter on INEEL Dissolved Calcine Slurries (INEEL/EXT-98-00749; August). Idaho Falls, Idaho: U.S. Department of Energy/Lockheed Martin Idaho Technologies Company.

McDaniel, E. W., and D. B. Delzer. 1988. FUETAP Concrete. Pp. 565-588 in *Radioactive Waste Forms for the Future*, W. Lutze and R. C. Ewing, eds. Amsterdam: North-Holland.

McNeese, L. E., J. B. Berry, G. E. Butterworth III, E. D. Collins, and T. H. Monk. 1988. Overall strategy and program plan for management of radioactively contaminated liquid wastes and transuranic sludges at the Oak Ridge National Laboratory (ORNL) (ORNL/TM-10757; December). Oak Ridge, TN: ORNL.

Miller, C. J., A. L. Olson, and C. K. Johnson. 1997. Cesium absorption from acidic solutions using ammonium molybdophosphate on a polyacrylonitrile support (AMP-PAN). *Separation Science and Technology*. 32:1-4.

Moore, J. G., H. W., Godbee, A. H. Kibbey, and D. S. Joy. 1975. Development of Cementitious Grouts for the Incorporation of Radioactive Wastes, Part I: Leach Studies (ORNL-4962). Oak Ridge, Tenn.: Oak Ridge National Laboratory.

Moore, J. G. 1981. A survey of concrete waste forms. Proceedings of a Conference on Alternative Nuclear Waste Forms and Interactions in Geologic Media (NTIS 194-216), J. G. Moore, L. A. Boatner, and G. C. Battle, eds. Springfield, VA: National Technical Information Service.

Murphy, J. A., L. Pincock, and I. N. Christiansen. 1995. ICPP Radioactive Liquid and Calcine Waste Technologies Evaluation Final Report and Recommendation (INEL-

94/0119; April). Idaho Falls, Idaho: U.S. Department of Energy/Idaho National Engineering and Environmental Laboratory.

Murray, R. F., D. W. Rhodes. 1962. Low-Temperature Polymorphic Transformations of Calcined Alumina. (Phillips Petroleum Company Atomic Energy Division, Contract AT (10-1)-205. [TID-4500, Ed. 17. IDO-14581. September] Idaho Falls, Idaho: U.S. Atomic Energy Commission.

National Research Council (NRC). 1965. The Radiochemistry of Plutonium. NAS-NS 3058. G. H. Coleman. ed. Clearinghouse for Federal Scientific and Technical Information, National Bureau of Standards, U.S. Department of Commerce, Springfield, Virginia.

NRC. 1999. An End State Methodology for Identifying Technology Needs for Environmental Management, with an Example from the Hanford Site Tanks. Washington, D.C.: National Academy Press.

Newby, B. J., J. H. Valentine, G. L. Cogburn, and C. M. Slansky. 1978. Calcination Flowsheet Development. Allied Chemical Corporation (INEEL TID-4500, R65; October). Idaho Falls, Idaho: U.S. Department of Energy.

Oak Ridge National Laboratory (ORNL). 1996. Integrated Data Base Report—1995: U.S. Spent Nuclear Fuel and Radioactive Waste Inventories, Projections, and Characteristics. (DOE/RW-0006, Rev. 12 – December). Washington, D.C.: U.S. Department of Energy.

ORNL. 1997. Integrated Data Base Report—1996: U.S. Spent Nuclear Fuel and Radioactive Waste Inventories, Projections, and Characteristics. (DOE/RW-0006, Rev. 13). Washington, D.C.: U.S. Department of Energy.

Office of Civilian Radioactive Waste Management. 1998. Civilian Radioactive Waste Management System Requirements Document Revision 4 (DOE/RW-0406, Rev4, May). Washington, D.C.: U.S. Department of Energy Office of Civilian Radioactive Waste Management.

Olson, A. L., W. W. Schulz, L. A. Burchfield, C. D. Carlson, J. L. Swanson, M. C. Thompson. 1993. Evaluation and Selection of Aqueous-Based Technology for Partitioning Radionuclides from ICPP Calcine (WINCO-1171; February). Idaho Falls, Idaho: U.S. Department of Energy/Westinghouse Idaho Nuclear Company, Inc.

Olson, A. L. 1997. Planned Cs Removal Technology (ALO-02-97; interdepartmental communication). Idaho Falls, Idaho: U.S. Department of Energy/Lockheed Martin Idaho Technologies Company.

Olson, A. L. 1998. Endstate Waste Form Requirements and Fates of Important Waste Constituents. Presentation given at the Idaho National Engineering and Environmental Laboratory (INEEL) Committee meeting on October 1-2, in Idaho Falls, Idaho.

Olson, A. L. 1998. Engineering Studies of Options. Presentation given at the INEEL Committee Meeting on August 18, at the Idaho Nuclear Technology and Engineering Center (INTEC) in Idaho Falls, Idaho.

Olson, A. L., and C. Barnes. 1998. TRU Separations Options. Presentation given at the INEEL Committee Meeting on August 18, at the Idaho Nuclear Technology and Engineering Center (INTEC) in Idaho Falls, Idaho.

Olson, A. L., and D. Taylor. 1998. Hot Isostatic Press (HIP) Waste Option. Presentation given at the NRC Committee's INEEL High-Level Waste Alternative Treatments Review on August 18, at the Idaho Nuclear Technology and Engineering Center (INTEC). Idaho Falls, Idaho.

Olson, A. L., and R. Kimmitt. 1998. Early Vitrification Option. Presentation given on August 18, at the NRC Committee's INEEL High-Level Waste Alternative Treatments

Review at the Idaho Nuclear Technology and Engineering Center (INTEC). Idaho Falls, Idaho.

Olson, A. L., and S. Losinski. 1998. Direct Cementitious Waste Option. Presentation given at the INEEL Committee Meeting on August 18, at the Idaho Nuclear Technology and Engineering Center in Idaho Falls, Idaho.

Organisation for Economic Co-operation and Development (OECD) Nuclear Energy Agency (NEA). 1995. The Environmental and Ethical Basis of Geological Disposal of Long-Lived Radioactive Wastes. Paris, France: OECD NEA.

Oversby, V. M., C. C. McPheeters, C. Degueldre, and J. M. Paratte. 1997. Control of civilian plutonium inventories using burning in a non-fertile fuel. *Journal of Nuclear Materials* 245:12-26.

Palmer, W. B., M. J. Beer, M. Cukars, J. P. Law, C. B. Millet, J. A. Murphy, J. A. Nenni, C. V. Park, J. I. Pruitt, E. C. Thiel, F. S. Ward, and J. Woodard. 1994. ICPP Tank Farm Systems Analysis (WINCO-1192; January). Washington, DC: Westinghouse Idaho Nuclear Company, Inc.

Palmer, W. B. 1996. HLW Treatment Alternatives Evaluation (WBP-29-96 Draft Report; August 16). Idaho Falls, Idaho: U.S. Department of Energy/Lockheed Martin Idaho Technologies Company.

Palmer, B. 1998. INTEC Waste Management. Presentation given at the National Research Council Committee's INEEL High-Level Waste Alternative Treatments Review meeting on August 17, at the Idaho Nuclear Technology and Engineering Center (INTEC). Idaho Falls, Idaho.

Palmer, B. 1998. HLW Baseline Plan, Accelerating Cleanup: Paths to Closure. Presentation given at the National Research Council Committee's INEEL High-Level Waste Alternative Treatments Review meeting on August 17, at the INTEC. Idaho Falls, Idaho.

Palmer, B., and A. L. Olson. 1998. Sodium Bearing Waste Treatment Study. Presentation given at the NRC Committee's INEEL High-Level Waste Alternative Treatments Review meeting on August 17, at INTEC. Idaho Falls, Idaho.

Palmer, W. B., C. B. Millet, M. D. Staiger, and F. S. Ward. 1998. ICPP Tank Farm Planning through 2012 (INEEL/EXT-98-00339; April). Idaho Falls, Idaho: U.S. Department of Energy/Lockheed Martin Idaho Technologies Company.

Parker, S. P. ed. 1994. Dictionary of Scientific and Technical Terms, Fifth Edition. New York McGraw-Hill.

Post, R. G. 1981. Independent Review Panel to Evaluate and Review Alternative Waste Forms to Immobilize the Idaho Chemical Processing Plant (ICPP) Calcined Waste, Final Report (ENICO-1088; December 1980). Idaho Falls, Idaho: U.S. Department of Energy Idaho Operations Office & ENICO.

Public Law (P. L.) 97-425 (1982) as amended by P. L. 100-203 (1987) and P. L. 102-486 (the Energy Policy Act of 1992).

Rawlins, J. K. 1998. Interim Storage Study Report (INEEL/EXT-97-01393; February). Idaho Falls, Idaho: U.S. Department of Energy/Lockheed Martin Idaho Technologies Company.

Raytheon Engineers & Constructors. 1994. Idaho Chemical Processing Plant Waste Immobilization Facility Feasibility Study Report, October 1994, Volume I— Feasibility Design Description (MS511.163; October). Idaho Falls, Idaho: U.S. Department of Energy Idaho Operations Office.

Rebish, K. J. 1993. Letter (May) from K. J. Rebish of Idaho National Engineering Laboratory to B. H. O'Brien, Manager, Pilot Plant Support and Development regarding multiple linear regression of nitrate versus both sodium plus potassium and excess calcium in selected pilot plant calcines. Idaho Falls, Idaho: U.S. Department of Energy.

Rechard, R., ed. 1995. Performance Assessment of the Direct Disposal in Unsaturated Tuff of Spent Nuclear Fuel and High-Level Waste Owned by U.S. Department of Energy, Volumes 1-3. Albuquerque, NM: Sandia National Laboratories.

Roberts, R., G. Choppin and J. Wild. 1986. The radiochemistry of uranium, neptunium and plutonium: an updating. Oak Ridge, Tenn.: Technical Information Center, Office of Scientific and Technical Information.

Ross, W. A., et al. 1983. Comparative leach testing of alternative transuranic waste forms. *American Ceramic Society Bulletin.* Pp. 1026-1035.

Russell, N. E., and A. D. Taylor. 1998. Hot Isostatic Press Waste Option Study Report (INEEL/EXT-97-01392; February). Idaho Falls, Idaho: U.S. Department of Energy/Lockheed Martin Idaho Technologies Company.

Russell, N. E., T. G. McDonald, J. Banaee, C. M. Barnes, L. W. Fish, S. J. Losinski, H. K. Peterson, J. W. Sterbentz and D. R. Wenzel. 1998. Waste Disposal Options Report (INEEL/EXT-97-01145; February). Idaho Falls, Idaho: U.S. Department of Energy/Lockheed Martin Idaho Technologies Company.

Sales, B. C., and L. A. Boatner. 1988. Lead-iron phosphate glass. Pp. 193-232 in *Radioactive Waste Forms for the Future*, W. Lutze and R. Ewing, eds. Amsterdam: North-Holland.

Schindler, R. E. 1974. Revised Design Criteria for ICPP Fourth Calcined Solids Storage Facility (September). Idaho Falls, Idaho: Allied Chemical Corporation, Idaho Chemical Programs, Operations Office.

Siemer, D. D., M. W. Grutzeck, and B. E. Scheetz. In Press. Paper on Comparison of Materials for Making Hydroceramic Waste Forms (abstract).

Siemer, D. D. 1995. Hot Isostatically Pressed Concrete as A Radwaste Form (abstract). *Environmental Issues and Waste Management Technologies* 61:657-664.

Siemer, D. D. 1995. Hot Isostatic Press (HIP) Vitrification of Radwaste Concretes. Pp. 403-410 in Materials Research Society Symposium Proceedings: Scientific Basis for Nuclear Waste Management XIX, Vol. 412. Boston, Mass.: Materials Research Society.

Siemer, D. D., D. M. Roy, M. W. Grutzeck, M. L. D. Gougar, and B. E. Scheetz. 1997. PCT Leach Tests of Hot Isostatically Pressed (HIPped) Zeolitic Concretes. Pp. 303-310 in Materials Research Society Symposium Proceedings: Scientific Basis for Nuclear Waste Management XX., Vol. 465. Boston, Mass.: Materials Research Society.

Siemer, D. D., M. W. Grutzeck, D. M. Roy, and B. E. Scheetz. 1998. Zeolite Waste Forms Synthesized From Sodium Bearing Waste and Metakaolinite. Waste Management '98. Session 36, Paper 2. Available at URL: http://www.wmsym.org/wm98/html/sess36/36-02/36-02.htm.

Slansky, C. M., ed. 1978. Technical Division Quarterly Progress Report October 1-December 1977, Idaho Chemical Programs (ICP-1141; February). Idaho Falls, Idaho: U.S. DOE/Allied Chemicals.

Slansky, C. M., B. R. Dickey, B. C. Musgrave, and R. L. Rohde. 1997. Technical Division Quarterly Progress Report July 1-September 30, 1997, Idaho Chemical Programs (ICP-1132; October). Idaho Falls, Idaho: U.S. Department of Energy/Allied Chemical.

Slaughterbeck, D. C., W. E. House, G. A. Freund, T. D. Enyeart, E. C. Benson, Jr., and K. D. Bulhman. 1995. Accident Assessments for Idaho National Engineering Laboratory Facilities. Science Applications International Corporation DOE/ID-10471. Idaho Falls, Idaho: U.S. Department of Energy.

Smith, H. D., E. O. Jones, A. J. Schmidt, A. H. Zacher, M. D. Brown, M. R. Elmore, and S. R. Gano. 1999. Denitration of High Nitrate Salts Using Reductants. Richland, Wash.: Pacific Northwest National Laboratory.

Sombret, C. 1998a. Letter (April 25) from C. Sombret to Tom Kiess of the National Research Council regarding some remarks about the fate of the INEEL calcined HLW. France.

Sombret, C. 1998b. E-mail (July 2) from C. Sombret to Tom Kiess of the National Research Council, regarding INEEL calcine. France.

Spaulding, B. C., R. A. Gavalya, M. M. Dahlmeir, L. C. Tuott, K. D. McAllister, K. C. DeCoria, S. P. Swanson, R. D. Adams, G. C. McCoy, R. J. Turk. 1998. ICCP Tank Farm Closure Study, Volumes I-III. Idaho Falls, Idaho: U.S. Department of Energy/Lockheed Martin Idaho Technologies Company.

Staples, B. A., G. S. Pomiak, and E. L. Wade. 1979. Properties of Radioactive Calcine Retrieved from the Second Calcined Solids Storage Facility at ICPP: Idaho Chemical Programs (UC-70—ICP-1189; March). Idaho Falls, Idaho: U.S. Department of Energy/Allied Chemical Corporation.

Staples B., H. Cole, and D. Pavlica. 1983. Properties of formula 127 glass prepared with radioactive zirconia calcine, p.125 in Scientific Basis for Nuclear Waste Management VI, ed. D. G. Brookins, Materials Research Society Symposium Proceedings V.15. Elsevier Science Publishing Co., Inc.

Staples, B. A. 1994. Composition of Calcines from H3 Campaign of the NWCF. Letter (October) to J. D. Herzog, Supervisor HLW Immobilization, INEL, Idaho Falls, Idaho: U.S. Department of Energy.

Staples, B. A., T. P. O'Holleran, D. A. Knecht, P. C. Kong, S. J. Johnson, K. Vinjamuri, B. A. Scholes, A. W. Erickson, S. M. Frank, M. K. Meyer, M. Hansen, E. L. Wood, and H. C. Wood. 1997. Preparation of Plutonium Waste Forms with ICPP Calcined High-Level Waste (INEEL/EXT-97-00496). Idaho Falls, Idaho: U.S. DOE/Lockheed Martin Idaho Technologies Company.

Staples, B. 1998. Twenty Years of INTEC Waste Form Development. Presentation given at the NRC Committee's INEEL High-Level Waste Alternative Treatments Review meeting on August 18, at the Idaho Nuclear Technology and Engineering Center (INTEC). Idaho Falls, Idaho.

Staples, B. A., D. K. Peeler, G. F. Piepel, J. D. Vienna, B. A. Scholes, and C. A. Musick. 1998. The Preparation and Characterization of INTEC HAW Phase 1 Composition Variation Study Glasses (INEEL/EXT-98-00970; September). Idaho Falls, Idaho: U.S. Department of Energy/Lockheed Martin Idaho Technologies Company.

TFA (Tanks Focus Area). 1998. TFA Technical Highlights: Period Ending October 30, 1998. U.S. Department of Energy/Pacific Northwest National Laboratory.

TFA. 1999. TFA Technical Highlights: Period Ending January 31, 1999. U.S. Department of Energy/Pacific Northwest National Laboratory.

Technical Peer Review Team. 1994. Report of the Technical Peer Review of the Idaho Chemical Processing Plant High-Level Waste Treatment Program. Technical Peer Review. Idaho Falls, Idaho: U.S. Department of Energy/Westinghouse Idaho Nuclear Company, Inc.

Thompson, M. In Press. Pretreatment Radionuclide Separations of Cs/Tc from Supernates (Document No. WSR-MS-98-00601). Savannah, Ga.: U.S. Department of Energy/Westinghouse.

Todd, T. A., K. N. Brewer, D. J. Wood, P. A. Tullock, L. G. Olson, and A. L. Olson. In Press. Evaluation and testing of inorganic ion exchange sorbents for the removal of

Cesium-137 from Idaho Chemical Processing Plant Acidic Tank Waste. (abstract). *Separation Science & Technology*.

Ullman, F., ed. 1985. Encyclopedia of Industrial Chemistry. Fifth Ed., 1st Rev., Vol. A1. VCH.

U.S. Department of Energy (U.S. DOE). 1985. An Evaluation of Commercial Repository Capacity for the Disposal of Defense High-Level Waste. Washington, D.C.: U.S. DOE.

U.S. DOE. 1987. Proposed Method for Assigning Metric Tons of Heavy Metal Values to Defense High-Level Waste Forms to be Disposed of in A Geologic Repository (DOE/RL-87-04 – August). Richland, WA: U.S. DOE.

U.S. DOE. 1996a. Waste Acceptance Criteria for the Waste Isolation Pilot Plant. Westinghouse Electric Corporation Nuclear Fuel Business Unit Waste Isolation Division. (DOE/WIPP-069 Revision 5; April). Washington, D.C.: DOE.

U.S. DOE. 1996b. Idaho National Engineering Laboratory Site Treatment Plan. (DOE/ID-10419; October). Idaho Falls, Idaho: U.S. DOE.

U.S. DOE. 1996c. Waste Acceptance Product Specifications (WAPS) For Vitrified High-Level Waste Forms EM-WAPS Rev. 02. U.S. DOE Office of Environmental Management.

U.S. DOE. 1997. Nevada Test Site Waste Acceptance Criteria (DOE/NV-335 NTSWAC, Revision 1; August). Reno, NV: U.S. DOE Nevada Operations Office Waste Management Division.

U.S. DOE Office of Environmental Management. 1997. Integrated Data Base Report-1996: U.S. Spent Nuclear Fuel and Radioactive Waste Inventories, Projections, and Characteristics. (DOE/RW-0006, Revision 13, December). Oak Ridge, Tenn.: Oak Ridge National Laboratory.

U.S. DOE. 1998a. FY1998 TFA Technical Midyear Review. Idaho Falls, Idaho: U.S. DOE Idaho Operations Office.

U.S. DOE Office of Environmental Management. 1998. Accelerating Cleanup: Paths to Closure (Draft. DOE/EM-0342; February). Washington, D.C.: U.S. DOE.

U.S. DOE. 1998b. Accelerating Clean up: Paths to Closure (DOE/EM-0362; June). Washington, D.C.: U.S. DOE Office of Environmental Management

U.S. DOE. 1998c. DOE Awards Contract. *Environmental Update* Summer (18).

U.S. DOE Alternatives Team. 1998. Process for Identifying Potential Alternatives for the INEEL High-Level Waste and Facilities Disposition Environmental Impact Statement (Predecisional Draft) DOE-ID 10627. Idaho Falls, Idaho: U.S. DOE.

Vinjamuri, K. 1991. Durability, Mechanical, and Thermal Properties of Experimental Glass-Ceramic Forms for Immobilizing ICPP High-Level Waste. High-Level Radioactive Waste Management. Proceedings of the Second International Conference Las Vegas, Nevada April 28-May 3, 1991. 708-713

Vinjamuri, K., S. V. Raman, D. A. Knecht, and J. D. Herzog. 1992. Waste Form Development for Immobilization of High Level Waste Calcine at the Idaho Chemical Processing Plant. Proceedings of the Third International Conference Las Vegas, Nevada April 12-16, 1992, High Level Radioactive Waste Management 2:1261-1271.

Vinjamuri, K. 1994. Glass-ceramic waste forms for immobilization of the fluorinel-sodium, alumina, and zirconia calcines stored at the Idaho chemical processing plant. *Ceramic Transactions* 45:3-12.

Vinjamuri, K. 1994. Soil-Based Glass-Ceramic Waste Forms for Immobilization of the Fluorinel/Sodium Calcined High-Level Waste Stored at the Idaho Chemical Processing Plant. Pp. 775-760 in Proceedings of the International Topical Meeting on Nuclear and Hazardous Waste Management—Spectrum '94 August 14-18, 1994. Atlanta, Ga: American Nuclear Society.

Vinjamuri, K. 1994. Effect of aluminum and silicon reactants and hip soak time on characteristics of glass-ceramic waste forms. *Ceramic Transactions* 45:265-275.

Vinjamuri, K. 1995a. Glass Waste Forms for the Na-bearing High Activity Waste Fractions. (INEL-95/0214; June). Idaho Falls, Idaho: U.S. Department of Energy/Lockheed Martin Idaho Technologies Company.

Vinjamuri, K. 1995b. Candidate glass-ceramic waste forms for immobilization of the calcines stored at the Idaho chemical processing plant. *Ceramic Transactions* 61:439-446.

Wichmann, T., N. Brooks, and M. Heiser. 1996. Regulatory Analysis and Proposed Path Forward for the Idaho National Engineering Laboratory High-Level Waste Program. (DOE/ID-10544, Revision 1; October). Idaho Falls, Idaho: U.S. Department of Energy/Lockheed Martin Idaho Technologies Company.

Wichmann, T. 1998. High-Level Waste Regulatory Framework and Issues. Presentation given at the NRC Committee's INEEL High-Level Waste Alternative Treatments Review meeting on August 18, at the Idaho Nuclear Technology and Engineering Center (INTEC). Idaho Falls, Idaho.

Wichmann, T. 1998. NAS Regulatory Framework. Presentation given at the NRC Committee's INEEL High-Level Waste Alternative Treatments Review meeting on October 1-2, at the Idaho U.S. Department of Energy Operations Office. Idaho Falls, Idaho.

Winograd, I. J. 1981. Radioactive Waste Disposal in Thick Unsaturated Zones. *Science* 212(4502):1457-1464.

Wood, D. J., J. D. Law, T. G. Garn, R. D. Tillotson, P. A. Tullock, and T. A. Todd. 1997. Development of the SREX Process for the Treatment of ICPP Liquid Wastes (INEEL/EXT-97/00831; October). Idaho Falls, Idaho: U.S. Department of Energy/Lockheed Martin Idaho Technologies Company.

Appendix B

Trip Reports of Technical Experts

Note: The INEEL committee invited several expert consultants to attend its first meeting and to write these trip reports. The opinions, findings, conclusions, and recommendations provided in these trip reports represent the views of the consultants and do not necessarily represent the views of the committee or the National Research Council.

DAVID CLARK
(University of Florida)

This report is based on oral presentations, written documents provided by INEEL and a tour of the INEEL facility. The INEEL's waste consists of about 5,300 m^3 of sodium-bearing liquid wastes (SBW) and 4,500 m^3 of calcined high-level wastes (HLW) in the form of solid granules. The SBW is stored in stainless steel tanks. The calcined wastes are stored in stainless steel bins designed to last for 500 years. Present storage facilities for both SBW and HLW are considered interim solutions. It is my understanding that DOE/INEEL is considering six options to permanently immobilize both types of wastes. These can be divided into two categories:

Separation Options

1. <u>Full separations, with grouting of low-activity waste (LAW) and vitrification of high-activity waste (HAW).</u> This option has two phases. In phase 1, the radionuclides, transuranics and fission products will be separated from the liquid SBW and subsequently vitrified and prepared for eventual geologic disposal off-site. The remaining liquid (LAW) will be immobilized in a grout and disposed of at INEEL. In phase 2, the calcine will be dissolved and the radionuclides will be separated and vitrified as in phase 1. Likewise, the remaining liquid LAW will be immobilized in grout.

2. <u>TRU separation/Class A grout + vitrification of Cs and Sr.</u> This option is similar to option 1. The major difference is that the transuranics will be separated from the fission products (Cs and Sr), placed in 55 gallon drums and sent to the Waste Isolation Pilot Plant (WIPP) for permanent disposal. The Cs and Sr will be mixed with an appropriate borosilicate glass frit, vitrified, and made ready for eventual geologic disposal.

3. <u>TRU separation/Class C grout</u>. This option is similar to option 2. The major difference is that the Cs and Sr will not be vitrified, but will be treated as LAW, immobilized with the other LAW in grout, and disposed of at INEEL.

Nonseparation options

4. <u>Early vitrification (also referred to a direct vitrification)</u>. In this option, the liquid SBW and solid calcine will be mixed with an appropriate borosilicate glass frit, vitrified, and prepared for eventual disposal in a geologic repository. Unlike options 1 through 3, the majority of the waste components will be treated like HAW.

5. <u>Direct cementitious.</u> This option is similar to option 4. However, the SBW will be calcined and the calcine will be immobilized in cement instead of in a borosilicate glass. Temperature requirements for this option are considerably lower than for vitrification.

6. <u>Hot Isostatic Pressing (HIP).</u> In this option, all of the SBW will be converted into calcine. All calcined wastes (SBW as well as existing calcine) will be mixed with an appropriate material (soil?) and consolidated under high temperature and pressure into a dense glass-ceramic. The resulting HAW glass-ceramic eventually will be sent to a geologic repository.

Although all six options have the potential to immobilize the wastes, option 1 (Murphy's Option 10, Murphy et al., 1995) involving full separations, grouting of the LAW and vitrification of the HAW appears to be the most favored by INEEL presently, based on the significant reduction in HAW volume that will have to go to a repository, on more even budgetary requirements over the next 20 years, and on total costs. From a total cost perspective, there is only a factor of about two to three difference between the most and least expensive options, and when storage and disposal are eliminated from the equation, the costs of all options are nearly the same.

The separations options will require a dissolution process that will put the solid calcine HAW wastes back into a less desirable liquid form. I have the same concern as expressed by Claude Sombret (1998) with separations. Why go to the trouble of making a calcine and then redissolving it? Although the volume of HAW will be considerably smaller using the separation approach, the total waste volume (HAW + LAW) most likely will be greater due to the additional volume of solution that must be used in the dissolution process. The separations approach appears to be driven primarily by the estimated (speculated) repository costs, which could change dramatically in the future. If there were additional reasons for separations, such as reclamation of isotopes for various applications, then this approach might become more reasonable. Whereas the options involving separations have merit and can significantly decrease the volume of HAW, it appears that the scale-up separation technology involving dissolution of the calcine has not been adequately demonstrated. I have a more specific concern with options 1 and 2. What happens if the concentrated HAW that is produced in the separations cannot be suitably immobilized in a borosilicate glass that can meet performance standards? If these two options have a high-priority, then much additional work needs to be performed to demonstrate the scale-up of the dissolution/separation process, and to demonstrate that the remaining radionuclides can be incorporated into an acceptable borosilicate glass.

Regarding the nonseparation options, both the early vitrification (frit + calcine) and direct cementitious (cement + calcine) appear attractive due to their simplicity and, in the case of vitrification, its proven track record. The major weaknesses in the direct cementitious option are the unknowns of the process scale-up in a remote facility and the requirement to demonstrate equivalency to borosilicate glass. There are two areas that require significant additional research/development even if early vitrification is the selected option. (1) There are differences in the calcine compositions within the different bins. Some thought needs to be given to blending these using a tank similar to the slurry melter evaporator (SME) at Defense Waste Processing Facility (DWPF) to make a more uniform waste stream (perhaps two blends; one high Al and one high Zr) (2) A database should be established from the literature and other sources of what already is known about vitrification of calcine in borosilicate glass. This information will be useful in developing one or two glass frit compositions, based on calcine compositions that will optimize waste loading while maintaining a low (1,150°C) processing temperature. Lastly, while the concept of a glass-ceramic appears promising, the use of HIP to achieve the product is too complex and scale-up too uncertain to pursue any further.

At the present time, a strong case for separations, based either on a technical or an economic basis, has not been made. Technically, the weak link in these options appears to be the dissolution of the calcine which still is in the exploratory stage of development. It is my opinion that the final decision for a waste form has to be based on the technical merits/issues such as processing simplicity, flexibility of the process, process maturity, and product performance and acceptance. Early vitrification (i.e., nonseparations) appears to be the best option when all of these factors are considered. However, my recommendation is for the INEEL staff to reduce the number of options to two through careful and detailed analysis of existing data. Additional research and development will need to be performed on both options before a final decision is made. If the early vitrification option is selected as one of the two, I would encourage the INEEL staff to develop strong interactions with the Savannah River Site and West Valley plants (and the French Cogema and Marcoule facilities) in order to take advantage of their "development" and "operations" expertise. Also, due to the time required to reach a final decision, it probably will be necessary to complete the calcination of the remaining 5,300 m^3 of liquid SBW.

Regardless of the option selected by INEEL, a more thorough analysis and characterization of the wastes will need to be performed. The product quality cannot be predicted or achieved if the feed stream is not well characterized.

EDWARD J. LAHODA
(Westinghouse)

General Observations

The INEEL effort is severely hampered by a lack of a good, representative sample of the waste tanks containing SBW and the calcine bins. Getting a good analysis of all the tanks and bins should be a top priority. For example:

1. Almost all options are predicated on obtaining an EPA exemption/delisting of the final waste form. It is doubtful that most of the organics listed in Wichmann et al. (1996) will be found if sampling is carried out. The presumed existence of these organics might also explain EPA's push to call the calciner an incinerator.

2. Thermal options (i.e., vitrification) enjoy an implicit advantage in that they are assumed to destroy the listed wastes allegedly contained in the SBW or calcine.

3. The results of any evaluation of options will enjoy a higher degree of confidence if their analysis is based on a sound waste feed composition.

4. Any potential problems due to the waste composition (i.e., Pu content and buildup) can be identified.

5. INEEL's laboratory and pilot-scale testing efforts are severely limited by the labeling of the waste as extremely hazardous. This label will likely be removed upon sampling.

6. The variability of the compositions of the calcines found in the bins is likely to affect how they are handled for processing. In addition, knowledge of this variability may require use of the unused bin set or even construction of a new bin set to help average out the range of calcine compositions.

The state of development for all the options considered is low. Therefore, the state of the technology development should not be a deciding factor at this time. This is not to say that the INEEL staff does not have technology strengths; they do and these strengths lie in solvent extraction and calcination. It is therefore understandable that the perceived best options at this point should involve the use of solvent extraction. This statement is based on the following observations:

1. The technical basis for the solvent extraction options are 1 liter, limited time tests.
2. The testing on other options has either been suspended for long periods of time or are proceeding at very low levels.

The cost estimates of the different options are probably no better than $\pm$ 100 percent for the processing portion of the options and should not be a determinant for the final option. Disposal costs are not likely meaningful numbers. This observation is based in part on:

1. Discounted values were not used. This would tend to minimize the effect of disposal costs.
2. The process designs used very conservative (and probably unrealistic) assumptions for operating factors. For instance, if one brought a melter on line, it is not clear why it would be run for only 180 days per year, especially if there was a standby already in line.
3. The process portions of the cost estimates were all within $\pm$ $1 billion dollars.

The choice of the final waste form should still be considered an open item. Since the long and arduous task of qualifying the final waste form has not even begun, it might be more reasonable to pick the waste form that is the easiest to produce. Therefore, consideration of a cement form should not be excluded at this early date.

The level of resources available to INEEL should be adequate to do the job.

1. In particular, an A&E with experience in costing vitrification, cementing, and solvent extraction processes at SRS or WVNS could be used for performing reasonable cost estimations. Utilization of SRS or WVNS operating personnel to help set reasonable operating assumptions should be encouraged. Could these personnel be borrowed from the other sites on a temporary basis to help perform the EIS?

2. Use of Hanford, SRS, and WVNS personnel to help develop and implement a sampling plan is encouraged. They have already gone through the agony.

3. Waste-form qualification and product quality performance should proceed using SRS as the model. This effort will likely require years to accomplish and much in the way of resources. Utilization of the SRS experience will hopefully reduce the level of expenditures required.

The concerns as to the institutional memory in terms of operations, design, etc., are well founded. This project will take at least 37 years to complete. This is much longer than a single generation (or maybe even two generations) of engineers is likely to last at this site. Therefore it is imperative that a project management approach be taken as to records generation, keeping, and retrieval now. Given today's technology, all reports, memos, letters and data should be put on a computer database that is backed up with all the normal controls and options (i.e., searching) that are available for any very valuable database. Maintaining written records is nice, but their limited searchability is a critical flaw.

Finally, I believe that the operability and maintainability of any process should be the primary basis for choosing a process. In this regard, simple is good.

In answer to the specific questions in the Statement of Task:

Are the set of treatment options chosen reasonable and are there others that need to be considered?

The options chosen are reasonable. Others that could be considered include:

1. Plasma melting as a replacement for a joule-heated furnace.
2. Glass/ceramic wasteforms not produced by HIP (i.e., Russian technology using thermite reactions).
3. Low pressure/temperature curing of cement type reactions.
4. Calcination of the remaining SBW and long-term storage on site. Although sugar has been the suggested organic additive to reduce the level of emitted NO_x and the residual NO_3 in the calcine, perhaps some other organic might be used which might have a reduced safety concern. One that comes to mind is acetic acid.
5. Utilization of one the current tanks to make it RCRA compliant might be considered. For instance, if there is a relatively clean tank available, perhaps another tank could be built inside of it. This would save the time and cost of building a brand new tank and of obtaining the required permit.

Are the assumptions, criteria, and methodology used reasonable?

The approach is reasonable. As per the above comments, I think that the process assumptions and costing portions of the analysis can be improved.

Are the environmental and technology risks reasonable?

1. The first priority should be the rendering of the residual SBW liquids to calcine as quickly as possible. Although this may slightly increase overall costs, this will result in a safer site for the long term until the program is completed over the next 30 years.

2. At this point in their development, I do not think that one can differentiate the identified technologies on either a safety or technology risk basis. For instance, until a complete flow sheet with mass and energy balances and at least a conceptual design are produced for each of the options, one cannot determine the costs, development needs or schedule in an adequate manner.

3. Risk assessments do not appear to have been run for any of the options. A risk assessment may be the means to provide the discipline necessary to carry out the EIS since it requires knowledge of the feeds, products, flowsheets, etc.

4. One point to note about the separations portions. Any solvent extraction process will likely add some level of organics to one or more of the streams. Will this be a concern for those options producing a TRU powder waste? Are any of the extractants or diluants considered hazardous, carcinogenic, potentially carcinogenic or mutagenic?

What is the feasibility of the treatment options within the regulatory framework and compliance agreements?

At this time, all appear to be feasible. Note, however, this statement is based on the assumption that reasonable resources are committed to doing the project. It is not wise to skimp during the initial phases where the final option is chosen or the laboratory development work is being done. The production of the EIS is likely to be the cheapest portion of the whole task and will set the course for the whole project to which very much larger sums of money will be committed. The next smallest cost will be the technology development effort. Extensive bench scale and pilot testing will save far more money than their cost.

K. K. SIVASANKARA PILLAY
(Los Alamos National Laboratory)

My participation in the National Academy of Sciences' assessment of the INEEL HLW alternative treatments leads me to provide the following observations and comments.

1. The range of technology alternatives proposed for the long-term management of HLW at INEEL is limited, and all options identified to be included in the upcoming EIS have high risks and high costs. None of these technologies is new or unique. These technologies have been extensively examined under the auspices of the ERDA/DOE in the past and those evaluations could be valuable to INEEL.

Prior to the passage of the Nuclear Waste Policy Act (NWPA) of 1982 by the U.S. Congress, there were extensive examinations of waste forms for the stabilization and geologic disposal of liquid HLW from spent fuel reprocessing. The DOE decision to choose glass vitrification technology for immobilization of HLW at Savannah River and West Valley was reached after considerable debate. ERDA/DOE reports published during that period by the Battelle Memorial Institute in Columbus, Ohio on these discussions would be valuable to this exercise.

2. Waste characterization data available to the NAS were mostly deduced from flow-sheet materials balancing, rather than from sampling and chemical analysis. As a result, considerable uncertainties exist in the viability of technology options proposed to meet the regulatory requirements. Identification of both safety and special nuclear material safeguards issues require realistic data on waste form compositions.

3. There is sufficient expertise at INEEL to conduct dissolution and separation of actinides and fission products from fuel materials of various compositions and irradiation history. However, at present there is very limited expertise and capabilities at INEEL to stabilize and manage the separated fractions for geologic disposal.

4. Adapting specific technologies for the treatment of HLW forms at INEEL will require experiments with INEEL waste materials, technology developments, and demonstration before determining the feasibility of alternatives. Consultation with experts on vitrification at SRS, PNNL, and West Valley are appropriate at this time.

5. From a first-hand knowledge of the new safety culture at DOE sites and the audits and assessments by DOE/EH, and the Defense Board, it is prudent to avoid wasting resources on technologies involving simultaneous use of high-temperature and high-pressure in processing highly radioactive materials in the canyons.

6. Although vitrification was the technology chosen by DOE in the 1980s for the solidification of liquid HLW at Savannah River and West Valley, a pragmatic alternative often proposed is the well-known waste cement technology. For serious consideration of this alternative, the performance requirements and comparisons to criteria developed for glass would need to be investigated if the glass option were not pursued as the baseline for HLW stabilization.

7. Recognizing the existence of negotiated agreements between DOE and the state of Idaho and the mandates of the court order of 1995, it is still prudent to consider the following simple, elegant, and relatively safe management strategy for the calcine waste forms at INEEL. The obligations of the agreement and mandate mentioned above have sufficient room for further negotiation to accommodate such an alternative.

A preferred alternative for the long-term management of existing HLWs at INEEL should be to convert all liquid wastes into calcine using the NWCF and leave the calcined wastes in the silos for several (10 to 20) decades. The natural decay of radioactive nuclides will allow a rather simple and safe disposition option of the long-lived radionuclides and RCRA constituents at the end of that period. The design life of the calcine bins is 500 years and they are sure to last without any major problems during the 10 to 20 decade period. This alternative is not the same as the "no-action alternative" being proposed in the EIS.

It should also be recognized that the waste forms created at INEEL as a result of reprocessing the calcines may not be accepted (or acceptable) at the WIPP or the Yucca Mountain facility, if and when they become operational. The consequence of the newly created waste form at INEEL would be worse than the worst scenarios associated with the storage of calcines in the silos for many decades.

The success of the above option will depend on a carefully formulated long-term financing plan to create an incentive for future generations to monitor, protect, and maintain the safety of these silos in good condition.

The reason for removing the calcines from Idaho to protect the Snake River aquifer is not quite justifiable considering the safe configuration of the calcine and the history of its performance in current storage silos. It would be more appropriate to direct attention and available resources to remove the infamous "pit-9" and similar pits from INEEL where solid wastes from Rocky Flats were dumped into open pits. A responsible resource management scenario would also choose the qualified "leave in place" option proposed here for the HLW calcines at INEEL.

8. So far, INEEL had very little problems with the HLW generated during the past 5 decades when compared with other three sites in the United States. At the same time, it should be recognized that privatization efforts within DOE complex have not been great successes. An objective reexamination of the real value added by privatization to INEEL HLW processing is appropriate well before such efforts are initiated.

9. Other issues that were not properly addressed and those that call for near-term examination include:

- a clear definition of the HLWs and quantities of such material at INEEL;
- the commingling of civilian and defense waste streams that may have impact on the acceptance of wastes at the WIPP site;
- the fissile nuclide composition of wastes and its significance to criticality safety and nuclear material safeguards;
- a management plan for the RCRA elements, such as Pb, Hg, Cd, etc., in the calcine and their end states (It is estimated that nearly 10 tons of mercury were used to catalyze the dissolution of spent fuels at INEEL. Establishing a material balance for this important RCRA metal as well as others before starting further processing of calcines is essential.);
- a strategy to manage long-lived radionuclides such as ^{237}Np, various Pu isotopes, ^{129}I, ^{99}Tc, ^{59}Ni, etc., that are present in the waste forms at INEEL; and
- an integrated examination of the HLW issues by the three Divisions of DOE (DP, EM, and MD) to arrive at a common strategy to manage the wastes within the DOE complex, including INEEL.

JOHN H. ROECKER
(Westinghouse Hanford Company, *consultant*)

The author of this trip report participated as a technical expert for the National Research Council Committee on INEEL HLW Alternative Treatments during its review conducted August 17-19, 1998, at Idaho Falls, Idaho. In keeping with the author's area of technical expertise these comments are focused on systems engineering and program management issues.

Systems Engineering

The application of aerospace systems engineering methods to nuclear waste processing is oftentimes made either unnecessarily complex, making it almost useless, or is simplified to such an extent that necessary steps are not performed to a sufficient degree. Considering the information presented, it appears that the INEEL alternative study focused primarily on the alternatives step.

In the application of systems engineering methods to an alternative treatment study, such as the INEEL study, four distinct steps must be performed to produce a product that will withstand critical scrutiny. These steps are as follows:

1. *Identification and definition of the initial conditions.* In the case of nuclear waste processing this will consist primarily of identifying and understanding the chemical and physical properties of the waste material, but should also included clear identification of any social, cultural, economic, and regulatory issues or conditions that may impact the processing scenarios.

2. *Identification and specification of a range of waste disposal end states.* This will consist of not only a set of chemical and physical characteristics that the final waste product must fulfill, but again, as in the initial conditions step, social, cultural, and economic (e.g., life-cycle cost) requirements must be defined. It should also be pointed out that it is essential that a range of end states be defined so a spectrum, not just a single set, of disposal scenarios will be examined.

3. *Definition of alternative processing functional scenarios.* Several sets of functional flowsheets that define various process combinations that will transform the waste from its initial condition into the desired end state need to be developed. All possible combinations of processing functions and all alternative technologies for a specific function should be identified initially. A screening review can be used to eliminate infeasible and impractical scenarios. The screening review should be documented so that review groups, such as this committee, can readily see why a particular scenario has been eliminated.

4. *Test alternative scenarios for performance against end-state requirements.* Simplified or screening performance assessments should be prepared to provide a measure of how well a processing alternative scenario performs. A few critical radioisotopes and hazardous chemicals can be utilized to develop a high-level performance assessment.

The following specific observations are made with regard to the INEEL systems engineering study:

1. Considering the information provided, it appears that the INEEL effort focused on step 3 of the systems engineering process and that steps 1, 2, and 4 either have not been performed or are incomplete.

2. With regard to the initial conditions, the chemical data presented for the calcine waste was based on calculated flowsheet information. This data needs to be substantiated by analytical laboratory sample results.

3. There is a considerable degree of uncertainty with regard to the quality of the chemical data presented in document (Garcia, 1997). One table was found to contain erroneous data. It was also stated by one of the study participants that the data in (Garcia, 1997) was not used in his study, but rather another set that was documented in the study report. This demonstrates a real need to establish a single source of certified waste characterization data, under configuration management, that is used by all participants.

4. End states need to be specified in terms of specific technical requirements that can be used by chemical process design engineers for development of alternative disposal scenarios. The requirements need to be quantitative not qualitative. Such requirements as "road ready" are not useful to the designer or the reviewer. The specific technical requirements to make the waste "road ready," if indeed that is the desired end-state, need to be defined, documented, and approved by the appropriate participants. When information does not exist, an enabling assumption should be used and a specific plan for resolving the open item defined. The level of detail in the end-state specifications will naturally become more developed and specific as the program matures.

5. A range of end states, not just those that satisfy the baseline, need to be defined so that both the decision maker and regulator will have a full spectrum of alternative disposal scenarios to consider. Expenditures of large sums of resources for little, if any, risk reductions are not worthy endeavors when less costly alternatives with acceptable environmental risks are available.

6. Several waste disposal processing scenarios have been evaluated by INEEL. However, without seeing a clearly defined range of end-states for the waste, it is probable that all practical alternative scenarios have not been considered.

7. Specific technologies familiar to INEEL personnel have been selected for the alternative disposal scenarios evaluated. However, it is not clear what consideration, if any, has been given to alternative technologies for a specific processing function. A range of technologies for each processing function needs to be evaluated to ensure that the most cost effective and most appropriate technology is selected.

8. There was no evidence presented that any screening-level performance assessment work has been performed to date. This is necessary so poor-performing disposal scenarios can be eliminated while work is continued on those alternatives that produce acceptable environmental risks.

Management Issues

1. Aggressive plans of action for resolving the "incidental waste" and "equivalency" issues are needed. If an unfavorable ruling is obtained on either of these issues, several current planning assumptions may be adversely impacted which could in turn affect the alternative selection. Timely resolution of these issues is necessary so the alternative selection decision process can proceed with certainty. Also, considerable effort by INEEL personnel may be required to provide sufficient information to the regulators. This effort needs to be incorporated into the INEEL program planning.

2. The continued use of the existing stainless steel tanks as an alternative to building costly new tanks needs to be evaluated. It appears that the existing tanks are sound and would continue so for some time in the future. An evaluation to determine whether the environmental risk from continued use warrants the considerable expenditure required.

3. Some top-level schedule evaluation for each alternative evaluated is necessary. Simply stating end dates does not provide much insight for the decision-maker into the schedule feasibility of an alternative. This is important for all alternatives, but particularly those requiring extensive technology development.

4. There was not much information regarding maintainability, operability, flexibility, and reliability presented. These items have to be factors considered in selecting a technology alternative. There are several examples throughout the industry of failed plants and/or processes because these items were not adequately considered.

5. The current assumption by INEEL personnel is that the facility(ies) will be privatized rather than government-owned contractor-operated (GOCO). However, most, if not all, of the information presented was developed using a GOCO approach. A privatization approach may well change the technical alternative, unless it is dictated by the request for proposal (RFP), and it will certainly change the cost profiles. If the technical approach is dictated by the RFP, it may unduly impact the bidding process. A determination on the privatization vs. contractor issue needs to be made and a consistent set of planning assumptions issued and utilized in the alternative selection process.

[Note: A detailed discussion of the application of the systems engineering methodology to a nuclear waste disposal program can be found in the National Research Council (NRC) publication *An End State Methodology for Identifying Technology Needs for Environmental Management: With an Example from the Hanford Site Tanks* (NRC, 1999)].

ERNEST RUPPE
(E.I. DuPont, *retired*)

Quality of the presentations and tours associated with the HLW disposal situation at INEEL was uniformly high and assisted the technical experts significantly in assessing the technology and situation.

In general, the problems at the site, though not simple, are in many ways more straightforward than at many other DOE sites:

- clear liquid wastes,
- sound tanks that do not leak, and
- calcine that is an excellent, handleable waste form for the interim.

Though it does not meet the present legal agreements, retention of all the waste at the site in this form in the well-designed bins provided for it is a solution which has a great deal of merit, obviating the personnel and transportation exposure and expense associated with any of the other plans.

Sorting out a preferred alternate from the EIS will of necessity be strongly influenced by the relative state of development of the options, some of which have lain dormant for a long time, presumably due to lack of funding.

There are several problems common to all of the alternatives:

- All have regulatory problems of one kind or another, but they have in common the need for delisting of RCRA components. All could founder on these. It appears that an aggressive approach to this problem, probably by DOE HQ with EPA, is needed. The regulators should all want to facilitate safe disposal of the waste, which will require the delisting. Hand in hand must go an aggressive sampling and analysis program, utilizing methods developed at other sites. This is necessary both for flowsheet development and to support the delisting effort, which will not have a high probability of success with the current kitchen-sink list.
- All alternatives need performance assessment and end-state analysis work. Wasteform qualification criteria/equivalence to existing vitrified products will take a long time based on Savannah River experience. It should be undertaken promptly for any alternatives to be actively pursued.
- The fate of low concentration but long-lived and active actinides in the wastes (technetium, ruthenium, and plutonium) must be determined.

Looking at the particulars of some of the processes presented:

Full Separations/Baseline Case

This is similar to work done elsewhere, and the separations steps are most familiar to Idaho personnel. Calcining the remaining SBW and redissolution of all the calcine followed by the vitrification is clearly a very expensive process. It appears to be the most developed option of those presented, but much development effort is still needed. Though some demonstration on the 1-kg scale has been made, extended runs of the liquid-liquid contactor system using real materials must be done to determine whether there are problems from buildup over extended recycle. Vitrification development is still at a very early stage, and, based again on Savannah River experience, will require a major effort.

Other options are all at a lower state of development because they are, in general, unique to INEEL and apparently are not active due to funding/staff limitations. Extensive development work is needed to seriously consider any. All have the common problems mentioned above. Though no staffing tables were presented, the overall impression was left that this work was not likely to be funded in the short-term future.

Early Vitrification Option

This option is attractive because it avoids the redissolution and separations of the first alternative, but again much demonstration is needed, including developing, characterizing, and qualifying the waste form. The volume downside will be discussed later in connection with costs. The waste form is most similar to "conventional."

Direct Cementitious Option

Direct grouting is attractive, except for the extreme volume downside attributed to it. The process cost appears very high, with the reasons not obvious.

TRU Separations

There are many known technologies here but also new ones to be developed. The option has not only the delisting problem common to all options but also the HLW reclassification, which is at least semantic but will be viewed critically. WIPP acceptance will also need to be developed.

HIP

Though HIP is used elsewhere for other purposes, it is difficult to envision hot cell operation of this complex technology at large physical size and 20,000 psi, 1050°C.

Based on experience in both high pressure and canyon operations, major practical and safety problems are apparent. It would be a very difficult development, and therefore low priority is recommended.

The glass-ceramic form, however is quite attractive. Are not other routes to it developable?

Finally, looking at the cost summary presented, volume and disposal costs drive these totals completely. The range of process costs shown is essentially the same within the limits of their accuracy. Thus the cost comparison hangs on use of the proper disposal cost value. This could range from incremental cost (considering previous expenditures, a sunk cost) to full cost. Alternatively if the waste is to go to a second repository, then replacement cost is appropriate. Further study, discussion, and agreement on this is needed.

Overall, the common problems noted at the beginning need to be pursued vigorously no matter which alternative is selected from the EIS.

BARRY SCHEETZ
(The Pennsylvania State University)

Introductory Comments

The visit to INEEL was very enlightening and indeed very useful in expanding my background knowledge of the site, facilities, and progress that has been made. The personnel at LITMCO and Idaho DOE should be complemented on their efforts in providing us with data and information that will be invaluable in our process for making recommendations.

However, more questions were raised during the presentations than could be answered within the time frame of the presentations and allocated discussion time. Many of these questions concerned fundamental assumptions that were made and the detailed rational behind the assumptions, for example, the issue of dioxane in the off-gas stream of the calciner. No clear explanation was given for the assumptions that it would be(is) present in the full-scale calcination process, especially since we were informed that no off-gas characterization has been performed. The consequences of proceeding with remedial actions based upon these assumptions are significant, not only in terms of project costs but also in the selection of a process flow sheet as well as regulatory involvement. Similarly, the assumption that all of the RCRA listed wastes that came onto site can be identified in the calcine, significantly and detrimentally influences the regulatory environment within which the calciner will continue to operate. The fear is that we heard what we were supposed to hear.

Discussion

In the early 1950s when the operators of INEEL made the decision not to commit to neutralization of their process wastes but to go with volume reduction via calcination, they made the most enlightened decision of all of the Atomic Energy Commission's facilities. Today some 40+ years later, their action still stand as a testament to good scientific reasoning; for instead of setting here today with a legacy of 56,000,000 gallons of corrosive, aqueous, mixed salt waste, we have 5,000+ m^3 of flowable, dry, granular, non-corrosive solids.

Separations

This approach is based upon a 1995 programmatic decision with its assumptions of for-cost and waste volume reduction based on Yucca Mountain National Nuclear Waste Repository, even though the waste from this project is designated to go to the second national repository. From the data present to this review committee, it appears that the desire to reduce the volume of HLW from 5,000+ m^3 to 210 m^3, irrespective of the consequences of this action, has driven the decision to justify this course of action.

It is apparent that the process technology is based upon strengths of the INEEL (separations). However, data that we received were lacking in the foresight to determine the fate of volatile hazardous and radioactive species such as Tc, Ru, and I in the proposed process; as if the technology was first selected and the process made to fit the technology. It appears that the data presented to us came from multiple sources with sometimes conflicting and, certainly, variable quality. Further progress will be impaired by lack of detailed knowledge of the compositions of the wastes; most information is currently based only on flowsheet calculations.

No discussion was provided to describe the nature of the projected 32,000 m^3 of "grout" that would contain the LLW residue from the separations process, nor the 70 tons of Cd and 12 tons of Hg that it would contain. No mention of the enormous volumes of nitric acid that would be required to take the calcine back into solution nor to its disposition were made. Although concern for the Snake River aquifer was presented as a significant motivating force for the removal of HLW, it should also be a motivating force for the removal of heavy metals as well. We have been presented with no information regarding the regulatory implications or stewardship requirements of building and operating a mixed-waste repository (landfill) that will remain within the state of Idaho. Are we to assume that the state of Idaho is aware of the implications of the construction and maintenance of such a LLW repository and concur with this action? We can only be left with the impression that these major considerations are in their infancy in development.

Furthermore, the separate operational steps associated with the proposed process is more complex (dissolution + separations + vitrification + grouting) than any of the proposed alternative approaches presented to us. The doctrine, "keep it simple" should be practiced.

Nonseparation Approaches

From these discussions, several very important statements were made. First, the fact that it is readily possible to homogenize the calcine immediately simplifies all alternatives processing. Instead of dealing with multiple, different calcines, only one will need to be addressed which should greatly simplify process designs. Second, we learned that up to 10 percent of SBW was already being mixed with existing calcine during its processing on a routine basis. Therefore, the further extension to include all denitration of the SBW and denitration of the residual nitrate in the calcine is readily feasible.

Three alternative approaches for the immobilization the SBW and calcine wastes were presented; but the alternatives were either dead for so long a period of time that their status is questionable, or the alternatives are so new that adequate funding has not been allocated for their development. In either case, we are left the conclusions that neither separations nor nonseparations options are well developed at this time.

Direct Vitrification

Direct formation into a glass was evaluated for approximately a 10-year period from 1976 to 1986, during which time the concept had advanced to the point that at least one full-scale glass log was produced. In the intervening year, West Valley and Savannah River have come on line with the production of full-scale radioactive glass waste forms. Scale-up and melter designs that have proven reliable can be implemented from the experience of these other sites. Development of frit compositions and evaluation of the resulting products are still needed especially in light of the stated ability to readily homogenize the calcine. Dealing with a single waste composition instead of as many as six variations greatly simplifies the processing flowsheets and minimizes the possibility of error.

Hot Isostatic Pressing into Glass-Ceramic

Hot Isostatic Pressing (HIP) has been developed at INEEL for nearly 2 decades. The data presented for this process appeared to be very positive, suggesting that both high waste loading and the formation of durable waste forms is readily achievable. The most significant disadvantage that appears to be of concern by other members of this review committee is the high operating pressure of the hot isostatic press. There was strong sentiment, expressed by several members of the committee who have direct operational experience, that such a system, with these very high pressures, would not be feasible to implement in a hot-cell environment. However, Argonne National Laboratory-West is actively pursuing the use of a hot isostatic press in their waste-form development. They have adequately demonstrated to another NAS review committee, on several occasions, the efficacy of placing a hot isostatic press, of up to 24 inches in diameter, into a hot-cell environment. Their schedule has advanced such that their objective is to place a full-scale hot isostatic press into a hot cell within 2 years.

Cementitious Waste Form

The data presented with respect to a cementitious waste form represent a minimal level of activity over the past 5 years. The cementitious waste form alternative that was presented to us in this review is not a conventional portland cement-based waste form. It is a cementitious material that is based on the alkaline activation of pozzolanic aluminosilicate materials. All development has been made with the availability of materials and the implementation of common industrial practices in mind. The cement hydrates to a C-A-S-H, x-ray amorphous solid, at room temperature rather than C-S-H that occurs with portland cement. When cured at temperatures in the range of 90°C to 175°C, the resulting product hydrates into a crystalline form, which possess a zeolite structure. Another unique aspect of these cements is that they have been designed to also posses the bulk chemistries that will allow them to be formed into glass-ceramics via HIP. HIP of this vitrifiable hydroceramic has been demonstrated to be achievable at 850°C and 2K psi to 5K psi while at the same time eliminating the difficulty in canister filling that is associated with the addition of dry powders to the HIP cans. The resulting product is a glass-ceramic not unlike that obtained in the HIP program. Waste loadings between 35 percent and 50 percent are achievable. It should be emphasized that the cementitious waste form before HIP can also meet durability requirements and may serve in its own right as a waste form while still maintaining the flexibility to be converted into a glass-ceramic (see Appendix A for a list of references).

The Fluor-Daniel report that was commissioned by INEEL to review this technology suggest that it could address the entire calcine problem in a much shorter time interval and at a cost that is a fraction of the $3+ billion earmarked for separations.

Recommendations:

Operating Assumptions

Some basic assumptions must be adopted (a) Since this waste will be going to the second generation national nuclear waste repository, for which no specifications have yet been established, "volume" should not be a driving issue. (b) Within the error of current estimates, there were substantially no differences in the cost associated with separations and all nonseparations approaches. (c) Separation of the high-level fraction of the waste is unnecessary, more complex, and should be discouraged. (d) Based on the second assumption, the calcines should be manipulated to ensure homogenization. (e) The current SBW and the calcine should be concurrently processed to remove the nitrate-load in the SBW and to reduce the residual nitrate in the calcines. (f) A processing approach should be adopted that meets minimal disposal requirements and is the simplest to operate. (g) Concerns over regulatory definitions of wastes should not drive the final decisions since these are likely to be modified.

Support for Alternatives Studies

With these fundamental assumptions in mind, several alternative approaches to handling the calcine now become viable but, because of the program's direction in the past 3 years, have not received the necessary funding to advance their state of development. In order to meet court-ordered deadlines, funding for these programs should be released immediately.

Ranking of Alternative Processes

Two previous independent commissions were charged with evaluating INEEL wastes with the objective of making recommendations as to types of waste forms that should be/could be implemented. In 1979, an NAS committee, chaired by Rustum Roy, concluded that immobilizing the calcine in a cementitious matrix was the most desirable approach to pursue. In 1981, another committee with much broader representation, chaired by Roy Post, evaluated 31 different waste forms and independently ranked them (Post, 1981). Again in the raw numerical ranking, a cementitious waste form, FUETAP developed at Oak Ridge National Laboratory in the mid- to late-1970s, was selected as the most desirable waste form.

The calcine, although designated HLW by its point of origin, is indeed a rather mild waste, at 40 watts/m^3, which would qualify as an "intermediate" level waste by U.K. standards. We can look to other organizations internationally (i.e., BNFL) and learn that wastes with this level of activity are routinely and successfully immobilized in a cementitious matrix.

The advantages of going to the process flowsheet discussed in this review are: (a) low temperature processing, (b) established "off-the-shelf" technology for processing concrete, (c) track record, (d) minimal op-steps, (e) flexibility to form hydroceramic initially which could differ processing into a glass-ceramic to some later date if desirable, and (f) the calcine waste becomes an integral component of the wasteform (i.e., soda, which is among the most difficult elements to immobilize, is necessary to initiate the hydration reaction).

The disadvantage of a hydroceramic at this stage is directly associated with the state of maturity of this specific formulation/process, although laboratory-scale development has proven successful. The process as it now stands has received virtually no attention. Funding for this process was less than $100 K from INEEL, in total.

The HIP process at INEEL has received a great deal of attention for nearly 2 decades and could complement the hydroceramic approach. HIPing a hydroceramic would resolve two of the major difficulties associated with the HIPing of powders by (a) eliminating the can-fill difficulties and (b) significantly reducing the P and T processing conditions. HIP is a viable alternative accepted by the NAS review committee for Argonne National Laboratory-West (ANL-W). A cooperative program should be established between INEEL and ANL-W, such that they would work together on the implementation of this technology.

Direct vitrification has also received a good deal of attention and in light of a homogenized feed, which would eliminate the need for multiple frits, should be evaluated.

Finally, a very promising option should be considered that would complete the calcination of the SBW and revert to permanent storage on site. With minor modifications (i.e., mounding earthworks around the above-ground silos for enhanced protections against the elements) to the existing bin sets, which already possesses a 500-year design life, they would form the basis for long-term storage/disposal. This alternative is not within the guidelines of court-ordered closures and shirks the government's responsibility to the state and people of Idaho and should only be considered as a "last resort" option.

Miscellaneous observations

It should also be emphasized that the resources to address the issue of alternatives to separation appear to be available at INEEL. To aid in the control of costs, outside contractors should be utilized where applicable. For example, we heard that the costs of modification of the calciner to meet clean air standards was a rate-limiting step and would cost perhaps a $100 million; while at the same time, a complete proposal for an entirely new calciner, specifically designed for handling SBW, including all of the emissions controls to met EPA requirements and at the total cost of $45 million, is available.

ANNE E. SMITH
(Charles River Associates)

Focus on Issues Related to Costs and Trade-Offs

The INEEL is in the process of developing cost and performance/risk estimates for a range of alternative processes for long-term management of the HAW at INEEL. To some degree, costs and cost-risk trade-offs will almost certainly be key considerations in the final decision, and this trip report focuses on those issues.

Although the official cost estimates for the final decision are not yet available, several preliminary and related cost estimates were cited among the presentations and reports that I reviewed. Some of these were quite detailed and some were "back of the envelope." It is important to recognize that these various cost estimates were never intended to be of comparable quality or mutually consistent. However, the differences among them have been useful for helping identify some specific cost-analysis issues that could benefit from general guidance. I have grouped the issues that I identified into the following five areas:

1. Discounting
2. Requirements for Completeness
3. Addressing Uncertainties
4. Role of Privatization in Costs
5. Assessing Trade-offs in a Multi-Attribute Setting

The comments that I provide below are not intended as criticism of any existing work. Rather, I am highlighting areas where there appeared to be a lack of consistent opinion or uncertainty about the most appropriate technical approach. I provide suggestions for approaches that are consistent with sound economics, and also adapted to highlight and address specific aspects of this waste management decision.

Discounting

The various cost estimates reviewed take different approaches to presenting treatment costs that occur over long periods of time. Some estimates "discount" costs and some do not. These inconsistencies can be explained by the different intended uses of the cost information in each document. The most important point here is that discounting *should be used* when the purpose of the cost results is to compare among the waste treatment options, with the ultimate goal of selecting a single alternative. Given that there were a number of questions at the August meetings regarding what discounting means, why we do it, and how we do it, I will provide a more detailed discussion of these points in this section.

A dollar to be spent in 10 years is usually viewed as less of a financial concern than a dollar to be spent today. There are many reasons that a given absolute amount of money declines in value over time. Most commonly one thinks of inflation. However, even if there is no inflation (or if the costs have already been adjusted to remove the effects of inflation), future dollars have less value, dollar for dollar, than current dollars. One reason is that people simply prefer choosing consumption now over consumption at some later date (when they may be less able to enjoy the experience, or may perhaps even die before then). Another very important reason is that if a dollar expenditure is put off, then the extra dollar available today may be invested in alternative activities that are productive (i.e., have financial returns). A productive investment means that *more* than one dollar of wealth will be available when the original dollar is to be expended at a later date. Hence, spending the dollar later will seem "less expensive" because the total wealth available to pay for it is larger.

"Discounting" is the process by which dollars spent in a future are converted into a comparable "present value." To compare the costs of alternatives that will involve expenditures at varying points in time, first all of the expenditures should be converted to a common "present value," then summed up to obtain an estimate of the total present value of the cost stream. If R is the annual discount rate, and H is the number of years into the future that expenditures on a project will occur, then

$$\text{Total present value of a project} = \sum_{T=1}^{T=H} (\text{Expenditure in Year t}) / (1+R)^t .$$

Some of the existing INEEL cost estimates use discounting. For example, the "Murphy report" (Murphy et al., 1995) uses an annual discount rate of 2.8 percent, applied to costs that do not include inflation. This contrasts to the budget estimates presented for the "Baseline" in the Ten-Year Plan (DOE, 1998). The latter report not only does not discount costs, but in fact escalates the on-going costs by 2.7 percent per year. How can these both be right? The answer lies in the specific use of the information in each report. The Ten-Year Plan is part of a budget document, intended for forecasting dollar amounts of future budget requirements. This "budget perspective" does not purport to provide a total social cost of the baseline strategy, and it does not compare the cost of the baseline to the costs of other options. It does not matter for purposes of congressional budget requests whether a future budget dollar is worth less than a current budget dollar. In fact, one can argue that a constant-dollar-amount budget allocation is easier to achieve than an allocation that is lower in present value, but increases over time in terms of dollars per year. It is also more important that the budget projection be presented in the dollar amounts that will actually be needed and requested. If these projections were presented in terms of discounted present values, this would only lead to confusion and increased probability of a failure to achieve the necessary funding when the future year comes. Thus, it makes sense that budget planning, *once an alternative has been selected*, be done without discounting future costs, and also that it provide for expected inflation in wages and cost of materials.

The budget perspective should be contrasted to the cost perspective, *which is the proper perspective to use in deciding among the HLW management options*. In this type of analysis, the goal should be to understand the total social costs of each alternative, regardless of whose budget the costs fall under or when the expenditures will be incurred. Thus, costs in future years should be deflated to present values and then summed up over the full time horizon. The time horizon to use should be long enough that it accounts for all the cost ramifications of each of the options. In this particular case, final disposal in most of the options is not expected until after 2065. Thus, the cost analysis for each option should extend about 75 years into the future.

The choice of discount rate to use varies with the context of the decision, and depends on the alternative uses of the funds if they are not spent today. Private corporations use "hurdle rates" that require that investments under consideration be assessed using discount rates from 10 to 25 percent per year. These hurdle rates include adders for private-entity concerns with risk, etc., and are not a good model for use in social investment decisions. For government investments, expenditures might be funded by Treasury Bills, which have an inflation-adjusted ("real") rate of return of about 3 percent per year. Alternatively, lower government expenditures might mean lower taxes, which in turn would mean greater consumption and investment by the private sector. This would suggest a discount rate between 3 and 8 percent.[1] Thus, the 2.8 percent discount rate used in the Murphy report is on the low end of the range that should be considered for the INEEL decision.

The present value can be quite sensitive to the choice of discount rate, so it will be reasonable to consider the effect on cost comparisons of discount rates over the range of 3 to 8 percent. A numerical example might be useful here. Consider two simplistic examples of management options. Option A involves an expenditure of $3 billion now and disposal costs of $0.5 billion 70 years later. Option B is cheaper up front ($0.5 billion now) but with a larger volume to dispose ($5 billion 70 years hence). Without considering discounting, the total costs of Option B appear to be $2 billion *more* than the costs of Option A. With a discount rate of 3 percent, a dollar to be spent 70 years from now is comparable to 17 cents to be spent today, and the present value of Option A is $3.1 billion, higher than that of Option B, which is $1.3 billion. The large differences in costs in the distant future may become insignificant in the comparison of the options. A potentially higher discount rate would only reinforce this finding. Thus, uncertainty about the precise discount rate to use may be less important to the INEEL decision than simply ensuring that some reasonable positive discount rate is used.

Requirements for Completeness and Consistency

Given that the INEEL HLW treatment decision is being made in the social interests, and not in the private interests of any specific entity or group of people, the comparison among the options should also account for all associated costs, regardless of the specific government budget that costs will fall under. Similarly, if there were any potential expenditure that would be borne by a non-DOE or nongovernment entity, this also should be encompassed in the cost estimate. Thus, the "cost perspective" may again result in quite different (higher) costs than one would obtain from a installation-specific "budget perspective" such as is reflected for INEEL in the Ten-Year Plan.

Two specific examples of completeness needs came to light during the August presentations: contingency costs and disposal costs.

Contingency costs are provisions for unforeseen costs or technical difficulties that cannot be anticipated in a standard engineering cost estimate. Although the specific form of the added cost is unknowable in advance, these are real costs, in that they are almost certain to occur due to one or more reasons. Accepted contingency factors are based on historical experience, and although there is some uncertainty in the precise magnitude required for any individual project, their expected impact on total costs is large: 20 to 30 percent typically. To exclude them from a cost estimate is to create an incomplete cost estimate.

Disposal costs are definitely part of the project costs that should be included in comparison of INEEL waste management options. These were excluded from the budget estimates of the Ten-Year Plan, but these must be included for the comparison among possible alterna-

---

[1] If inflation were to be built into the cost estimates, then the rate of inflation should also be added to the discount rate. For example, if the real discount rate is to be 5 percent, and the cost estimates have a provision for 3 percent inflation, then the present value calculation should be performed using 8 percent (i.e., 5 percent + 3 percent). If there is no difference between the economy-wide rate of inflation and the rate of inflation expected for the project in questions, then there is no need to build inflation into the cost estimates. Doing so will not change the present values, as the discount rate must also be adjusted upwards in an offsetting manner. Accounting for inflation does make sense for budget planning purposes, however.

tives. Several people at the meetings noted that the disposal costs are the real drivers in the total costs. For example, the last page of the second package of slides handed out at the August meeting shows a "preliminary cost summary." If the options on this slide are ordered in terms of "total costs," and also in terms of "disposal costs," the same ranking is obtained. However, there are reasons why disposal costs may not be as significant as this would suggest:

- First, this slide has not presented *present values* of costs. Even with a fairly low discount rate of about 3 percent, disposal costs would be discounted by a factor of 3 to 6, relative to the waste treatment costs.
- Second, the cost-per-ton assumption for disposal of the final waste form might be overstated. This assumption has been based on the total costs of getting the Yucca Mountain repository into functional status, divided by the tons of waste that can ultimately be disposed in Yucca Mountain. Since Yucca Mountain is the first such repository, it may have involved substantial start-up costs that would not be fully repeated with future repositories, possible expansions of capacity at Yucca Mountain, or future acceptable methods for HLW disposal. A member of the public who spoke at the August meetings argued that costs per ton for disposal of HAW should be based on the incremental cost necessary to dispose the incremental tons, and that these costs would be only $1,000/ton, a factor of 300 less than the unit disposal costs being used by DOE at this time. I agree that the cost per ton should be based on the incremental costs of disposing of HAW rather than the average cost based on the first increment represented by Yucca Mountain. I disagree, however, that this incremental cost will be as low as $1,000/ton. New repositories or new disposal capacity will require additional costs above and beyond the engineering cost of transporting and placing the waste canisters in a deep repository. Conceivably, future repository capacity may be more costly to set up and approve than Yucca Mountain has been. However, a better estimate of the disposal costs that attempts to eliminate any "one-time" aspects of the Yucca Mountain costs would be a good idea and could help define a range of cost estimates for the important disposal cost component.

Despite the uncertainty about their overall contribution to total costs, disposal costs *should* be included in the cost estimates for each alternative.

Consistency is required in cost estimates as well, for completeness. The costs (and feasibility) of some alternatives are less certain than for other alternatives. Costs may be higher than a point estimate that would be developed with current information if there are technical complications that cannot be easily surmounted. They may be lower than a current point estimate would indicate because processes may be better optimized. The cost estimate range will tend to be larger for less mature treatment technologies. Completeness demands that ranges of costs be provided to help normalize for the "maturity" of the various cost estimates.

Addressing Uncertainties

Uncertainty is obviously a key feature of cost estimates for unproven technologies and activities that are spread out over many decades. None of the cost analyses presented at the August meetings have effectively described the uncertainties in ways that can help in the decision making, and a better job will be required in this area. Uncertainty can be incorporated into an evaluation in manageable ways. Two key principles are (1) focus on uncertainties might affect the relative ranking of options (e.g., uncertainties that significantly affect the costs of some options but not all); and (2) explore significant key assumptions in a "what-if" mode.

In the first category, one might include uncertainties that are driven by significant differences technological maturity among the options, as described in the previous section. Similarly, uncertainties in disposal costs may be important in differentiating among options.

In the second category, there may be some assumptions that affect all of the cost estimates in a similar manner, but which may have dramatic effects on costs. One example that comes to mind is the assumption that DOE will successfully get a delisting of hazardous wastes. All of the cost estimates depend significantly on this single assumption, yet there appears to be a significant risk that delisting will be denied. A most informative element of the cost analysis would be an exploration of what would happen to the waste management options and decision if delisting were to be denied. What would the alternatives look like? How would this event alter the comparison among the options? Would one option suddenly appear much more advantageous? Would some options become anathema? This type of "what-if" analysis need not be highly quantitative, as long as the options are carefully scrutinized for pitfalls and advantages, and the implications of this assumption for costs are openly discussed.

Another important assumption appears to be that the designation of the calciner as a hazardous waste incinerator, with attendant upgrading costs and waste management delays, is immutable. I should think that this decision, which appears to be fairly poorly thought out by regulators, could be reopened if a good risk-risk comparison were made to support the discussion. An important question in my mind is how the costs and options being considered might be altered if the calciner could be run without major upgrading until the remaining liquid wastes are calcined.

An important feature of a good uncertainty analysis is that it highlights the most salient aspects of the decision and targets them for greater discussion and exploration. Doing so may lead to better alternatives generation. For example, in the August meetings, it became apparent that there are really two basic options, and that focusing too heavily on detailed comparisons among all of the waste treatment options may be missing a more important aspect of this decision, which is whether or not to treat and remove the calcined wastes at all.

Figure B-1 illustrates the nature of the INEEL decisions in terms of a cost and risk trade-off. (Risk is defined strictly in terms of environmental risks in Figure B-1.) A balloon illustrating the range of cost and risk uncertainties represents each of the options. "No action" has low but non-zero costs. The option to finish calcining the wastes and to store them permanently in the existing bins will probably give a substantial risk reduction, possibly with minor increases in costs. (Although there are costs to calcining, the life-cycle costs might be reduced due to lesser monitoring and maintenance needs compared to the "no action" alternative.) The next increment in action is conversion of the wastes into a stabilized form and off-site disposal. All of the technologies discussed at the August meetings are associated with this general type of action. As can be seen, they all appear to involve very large cost increases and probably minimal further environmental risk reductions. This figure highlights the basic fact that there may be too much attention being given to defining and analyzing all the conceivable "stabilize and remove" options and too little attention to intermediate alternatives. Some intermediate alternatives that could be considered might be (a) upgrades to the calcined wastes and/or the bins to make them lower-risk as a permanent configuration, and (b) waste vitrification, but with on-site permanent storage. Although these may not be considered politically or legally acceptable at this time, it is not possible to know whether these intermediate options deserve further regulatory consideration unless their cost and risk attributes are explored in more careful detail.

Figure B-1 gives the sense that there really is a cost-risk trade-off to decide on. However, there are more dimensions to risk than just the environmental. There are safety concerns associated with the treatment activities, and the safety risk almost certainly increases with the amount of treatment. Figure B-2 illustrates a possible cost-risk curve when risk is redefined to include both environmental and human safety risks. If information available today were summarized in this form (Figures B-1 and B-2 are hypothetical and purely meant to illustrate methods for thinking about trade-offs), the debate may be better targeted to finding "better" options, and this might lessen the importance of obtaining precise cost estimates for the waste treatment technologies. The differences among the treatment alternatives described in August may start to seem relatively minor compared to the basic questions of whether the attendant incremental risk reduction is worth the incremental cost, and whether there might be intermediate solutions providing more reasonable trade-offs.

Role of Privatization in Costs

A few times during the August meeting, the discussion came around to the possibility of this waste treatment activity being privatized. Privatization is assumed in the Ten-Year Plan for budget purposes, with an attendant assumption that costs will be reduced by 20 percent over the existing estimates. I do believe that privatization can lead to cost decreases over a non-privatized clean up. However, I do not believe that engineering cost estimates will be decreased by 20 percent, or any other arbitrary factor. Rather, privatization may only reduce the chance that actual costs will *exceed* the estimates.

I could go into some depth about why these factors should not be used. However, we should also note that these factors would apply to all of the options equally. An across-the-board scaling down of cost estimates will not affect the cost ranking of the alternative approaches. It is therefore not important to dwell on numerical assumptions. Rather, it should be noted that the benefits of privatization derive from the way that a carefully thought-out privatization scheme creates management flexibility and incentives to make good use of that flexibility. The longer privatization is delayed in the process of deciding how to best manage the INEEL wastes, the less the opportunities to gain from its potential benefits. The current plan assumes privatization will only occur *after* the particular waste treatment technology has been identified and developed. At that point many of the potential benefits of privatization may have been lost.

I therefore suggest that if privatization is to be attempted that DOE seriously consider the merits of speeding this process up. I also recommend that attempts to incorporate the effects of privatization into cost estimates be avoided: it won't improve the cost estimates, and it won't substantively affect the decisions among alternatives.

Assessing Trade-offs in a Multiattribute Setting

Everyone recognizes that neither costs, nor risks, nor cost-risk trade-offs alone will determine the best alternative. There are other attributes, sometimes less quantifiable, that people consider important as either leading terms or modifiers to the decision among alternatives. The "Murphy report" (Murphy et al., 1995) provides an assessment of alternatives that accounts for the role of multiple attributes and alternative values among different interested parties. This report is a good example of some of the things that can be done to help illuminate options that are robust across a range of criteria and a range of value systems, without resorting to the controversial and sometimes "black-box" approach known as "multi-attribute utility analysis" (MAUD).

I recommend that multiattribute summary presentations also be used for the final INEEL waste treatment decision, when the time comes to integrate all of the technology information. The approach initiated in the Murphy report can serve as a starting point. I also point to some additional tools and techniques that can help in visually representing the ways that different alternatives make trade-offs among multiple decision-relevant attributes. In the past, I assisted the U.S. EPA in developing and applying visual multiattribute assessment techniques to large, long-term decisions such as regional adaptation strategies for climate change. The approach was facilitated by a PC tool called TEAM (Tool for Environmental Assessment and Management), which EPA distributes. However, the concepts are simple, relying on visual techniques for representing relative and qualitative summaries of how various alternatives perform with respect to multiple different decision criteria. The usefulness of a PC tool is that it provides automatic access to alternative visual display formats and allows groups of users to explore the implications of alternative weighting of attributes in the visual display.

Figure B-1. Illustrative Example of Nature of Cost-Environmental Risk Trade-offs for INEEL High Level Waste Management Decision

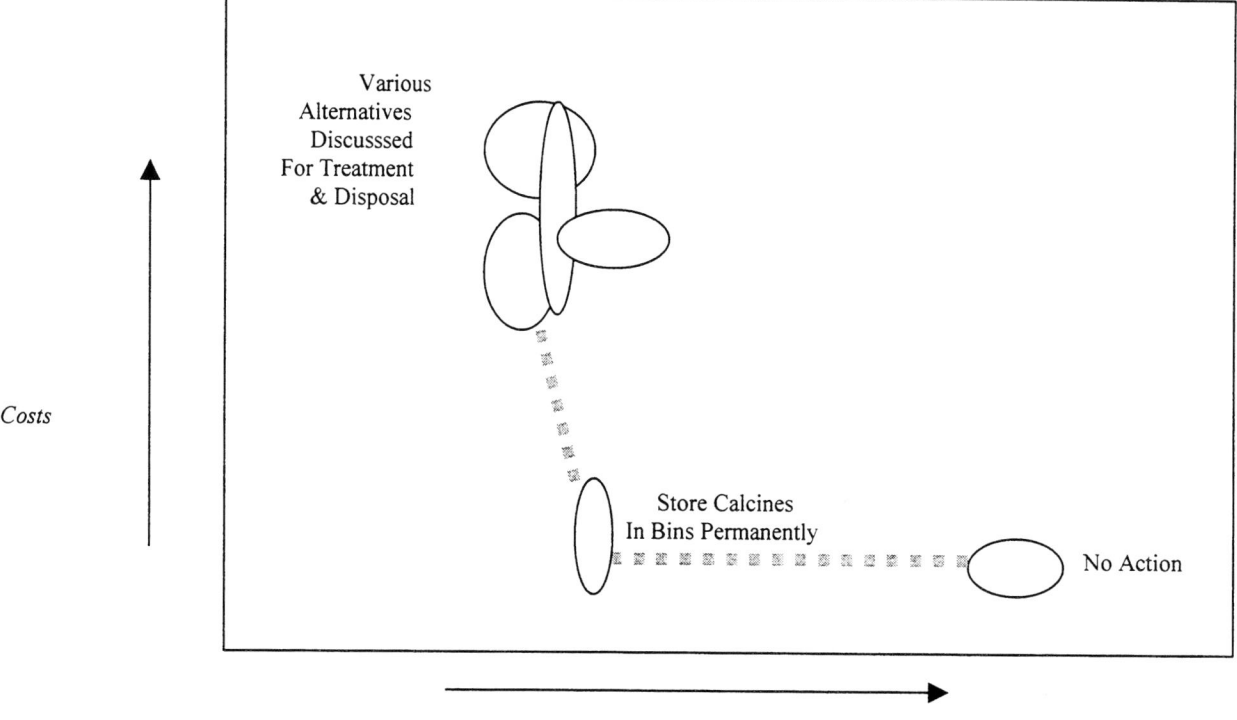

Figure B-2. Illustrative Example of Nature of Cost-Total Risk Trade-offs for INEEL High Level Waste Management Decision

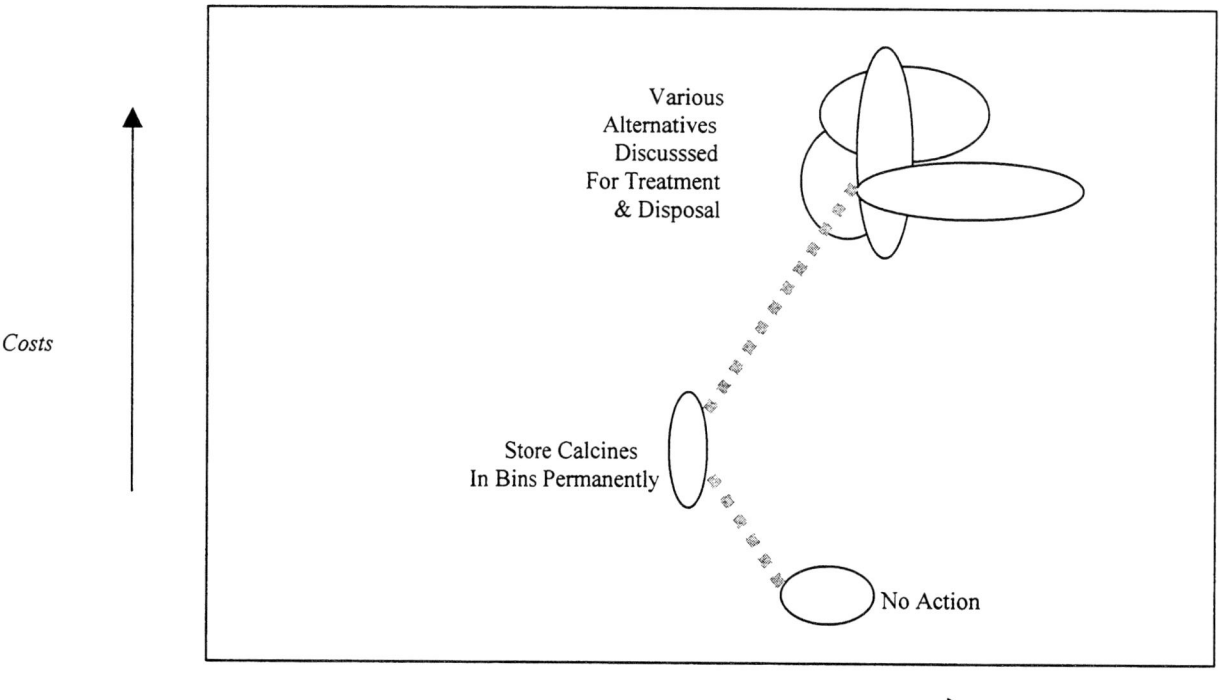

Appendix C

Biographical Sketches of Committee Members

Robert C. Forney (NAE), *Chair*, is a retired executive vice president, member of the board of directors, and member of the executive committee of E.I. Du Pont de Nemours & Co., Inc. During his almost 40-year career with du Pont, he held a wide variety of research, manufacturing, engineering, marketing, and general management positions, the first 27 years of which were in man-made fiber activities. He is a member of the National Academy of Engineering and serves on the board of several for-profit and not-for-profit organizations. Dr. Forney received his B.S. and Ph.D. in chemical engineering and his M.S. in industrial engineering, from Purdue University.

Edward Aitken is currently a consultant and retired manager of materials development with the General Electric Company. Dr. Aitken has over 35 years of experience in process technology, materials development and systems engineering applications. His expertise is in fabrication, product development and waste management of nuclear materials involving both ceramic and metal systems. Dr. Aitken is a fellow member of the American Ceramic Society, and the American Nuclear Society of which he is a past chairman of the Materials Science and Technology Division. He received a B.S. degree in chemistry from the University of California-Berkeley in 1950 and a Ph.D. degree in physical chemistry from the University of California-Los Angeles in 1954.

Robert Bertucio is an engineer and project manager with SCIENTECH, Inc. His expertise is in probabilistic risk assessment and methods development. Dr. Bertucio has 24 years experience in the area of performance, management, and review of probabilistic safety and risk assessment (PSA/PRA) and has managed PSA programs for the Department of Energy, the U.S. Nuclear Regulatory Commission, and national laboratories. He received a B.S. degree in aerospace engineering from Syracuse University, and M.S. and Ph.D. degrees in nuclear engineering from Carnegie-Mellon University.

David O. Campbell is a consultant and retired nuclear chemist from Oak Ridge National Laboratory (ORNL). His expertise includes nuclear fuel reprocessing, radioactive waste processing, solvent extraction of metals, and nuclear waste management. Prior to retirement, Dr. Campbell's research focused on improved methods for treatment of radioactive liquid wastes at the ORNL site. He is a member of the American Chemical Society and the American Nuclear Society (ANS). He is the recipient of the ANS Special Award for Advancements in Nuclear Technology in Response to Three Mile Island, author of numerous publications, and holds several patents. Dr. Campbell received a B.A. degree in mathematics and chemistry from the University of Kansas City (now University of Missouri at Kansas City) in 1947, and a Ph.D. in physical chemistry from the Illinois Institute of Technology, Chicago, in 1953.

Melvin S. Coops is currently a consultant to Lawrence Livermore National Laboratory (LLNL) and Los Alamos National Laboratory (LANL) and, through the University of Chicago, is a technical reviewer of the Nuclear Technology Program at Argonne National Laboratory. Mr. Coops experience in chemical separations of radioactive materials, actinide metallurgy, and remotely operated processing systems spans more than 40 years. He is an expert in nuclear fuel cycle chemistry using both aqueous methods and pyrochemical techniques. His experience includes separations chemistry of the fission product, lanthanide, and actinide elements, with a special interest in the chemistry, metallurgy, and nuclear properties of the isotopes of uranium, neptunium, plutonium, americium, and curium. Mr. Coops is retired from LLNL but continues to work there part-time. He has been a member of the American Nuclear Society since its inception. He holds a B.S. degree in chemistry from the University of California at Berkeley and a Sc.D. equivalent from LLNL.

Delbert E. Day is the Curators' Professor of Ceramic Engineering at the University of Missouri-Rolla and Senior Investigator of its Graduate Center for Materials Research. His teaching and research have dealt with the structure and properties of vitreous solids (glass) and solid waste recycling. He holds 38 U.S. and foreign patents dealing with glass microspheres, sealing glasses, ceramic dental materials, and oxynitride glasses, and he is a past president of the American Ceramics Society. Dr. Day received a B.S. in ceramic engineering from the Missouri School of Mines and Metallurgy (now University of Missouri-Rolla), and received an M.S. and a Ph.D. in ceramic technology from Pennsylvania State University.

P. Gary Eller is a project leader for the Advanced Technology Group of Los Alamos National Laboratory (LANL) and currently manages the research and development effort to provide safe stabilization and storage of nuclear materials under the DNFSB 94-1 program. His expertise is in actinide, environmental, and fluorine chemistry and applications to the nuclear fuel cycle. He has 25 years of research and development experience in nuclear materials and environmental issues, and 11 years of experience in the leadership and management of projects such as radioactive site cleanups and the remediation of nuclear materials. Dr. Eller is a member of the American Chemical Society and the American Nuclear Society and has a host of awards. He has more than 100 journal articles, 3 patents, numerous LANL reports, and several other publications. Dr. Eller received a B.S. degree in chemistry from West Virginia University in 1967 and a Ph.D. degree in inorganic chemistry from Ohio State University in 1971.

Rodney C. Ewing is a professor in the Department of Nuclear Engineering and Radiological Sciences at the University of Michigan and an adjunct professor there in the Department of Geological Sciences and at the University of Aarhus in Denmark. His expertise is in mineralogy and materials science, specifically, radiation effects in complex ceramic materials and long-term durability of radioactive waste forms. Dr. Ewing has conducted research in Sweden, France, Germany, Australia, and Japan, as well as the United States. He is a past president of the International Union of Materials Research Societies. He has served on several National Research Council committees. He also has served on the subcommittee on the Waste Isolation Pilot Plant for the Environmental Protection Agency's National Advisory Council on Environmental Policy and Technology. Dr. Ewing received both M.S. and Ph.D. degrees in geology from Stanford University in 1972 and 1974, respectively.

John M. Kerr is a consultant with Innovative Technologies, Inc., in Lynchburg, Virginia. Mr. Kerr retired from Babcock and Wilcox as an Advisory Engineer in 1995. His expertise is in ceramics and chemical engineering specializing in using ceramic bodies for waste disposal, ceramic fuel-metal compatibility studies, ceramic fuels development and development of fuels fabrication methods, nuclear ceramics, and nuclear fuel cycle studies. He is the author and co-author of several publications and articles relative to this area of materials research and development. He is a member and fellow of the American Ceramic Society and the American Materials Society. Mr. Kerr received a B.S. in ceramic engineering from the University of Illinois in 1956, and an M.B.A. from Lynchburg College in 1973.

Jean'ne M. Shreeve is the vice president for research at the University of Idaho and a professor of chemistry. Her expertise involves the synthesis, characterization and applications of fluoride compounds. She has an extensive background in studies relating to fluoride chemistry and high-temperature fluids. She is a member of the American Association for the Advancement of Science, the American Chemical Society, and the Royal Society of Chemistry. She has authored or co-authored over 290 refereed publications. Dr. Shreeve received a B.A. degree in chemistry from the University of Montana, a M.S. degree in analytical chemistry from the University of Minnesota, a Ph.D. degree in inorganic chemistry from the University of Washington-Seattle, and carried out postdoctoral work in fluorine chemistry at Cambridge University.

Minoru Tomozawa is a professor of materials science and engineering at Rensselaer Polytechnic Institute. Dr. Tomozawa joined the faculty at Rensselaer in 1969, after working for the Nippon Electric Company. He has published extensively in the area of glass science and edited several books on the subject. He is a past chair of the Glass and Optical Materials Division of the American Ceramic Society and a fellow in the American Ceramic Society. Professor Tomozawa's research interests are the structure and properties of glasses and glass-ceramics. His research aims to characterize the structural changes of such glass in a controlled manner and to alter materials properties in a desired manner, lead to glass of improved stability. Professor Tomozawa received a Ph.D. degree in materials science and metallurgy from the University of Pennsylvania in 1968.

Appendix D

Biographical Sketches of Technical Experts

David E. Clark has taught in the Department of Materials Sciences and Engineering at the University of Florida since 1978 and has been a full professor since 1986. He has conducted research programs in the areas of nuclear waste materials, environmental degradation of glass and ceramics, microwave processing, ceramics, sol-gel processing, coatings, glass, ceramic superconductors and self-propagating high-temperature synthesis that have supported approximately six graduate students each year. Dr. Clark is actively involved in professional organizations including the American Ceramic Society, the National Institute of Ceramic Engineers and the Materials Research Society. In each of these organizations, he has organized numerous symposia and served as primary editor for no less than ten symposia proceedings. He has well over 200 publications across the research areas listed above. In addition to serving as a technical expert for this review committee, he has chaired a technical review group that evaluated a compendium on nuclear waste glass for a program managed by Argonne National Laboratories for the Department of Energy (1992-94). Dr. Clark also served as a member of a peer review committee to assess and evaluate the glass-sampling program for the Defense Waste Processing Facility (DWPF) at Westinghouse Savannah River (1990-91).

Edward J. Lahoda has over 20 years of experience in process analysis, development, design, and field support. He is responsible for the technical developments of the Westinghouse soil washing and high-temperature thermal desorption technologies. He has chemical process design experience in processing chemical warfare agents, nuclear fuels, high- and low-level nuclear wastes and plasma processing of wastes and plasma production of specialty materials. He has provided field support to operating facilities including the Westinghouse incinerators, nuclear fuels production, steam generator maintenance, soil washing, and thermal desorption operations. He has served on committees at the Savannah River site addressing overall operation of the DWPF and test data validity for DWPF and chaired the ITP Chemistry Review Panel. Dr. Lahoda received B.S., M.S., and Ph.D. degrees in chemical engineering and an M.B.A. from the University of Pittsburgh. He is also a member of the American Institute of Chemical Engineers.

K. K. Sivasankara Pillay is a senior staff member at the Los Alamos National Laboratory. His 35 years of work experience in the nuclear science and technology fields include his tenure as a graduate faculty member at the Pennsylvania State University and as a researcher with considerable success in bringing nuclear technologies to beneficial uses. During the past two decades, he has actively participated in various aspects of waste management associated with fissile materials production activities. He is well conversant with issues relevant to excess fissile materials and waste management with the U.S. weapons complex. At Los Alamos, he spent the first ten years in areas of nuclear materials management and international safeguards. He is presently with the plutonium facility at Los Alamos helping the facility to achieve its waste minimization goals through technology innovations. Dr. Pillay is a fellow of the American Nuclear Society and the American Institute of Chemists. He is a member of the American Chemical Society, American Association for the Advancement of Science, the Institute of Nuclear Materials Management, the Health Physics Society, and the ASTM. He is the author of over 170 publications in the open literature. Until recently, he was an associate editor of the *Journal of Radioanalytical and Nuclear Chemistry*. Dr. Pillay received his B.Sc.(Hons) and M.Sc. degrees in chemistry and physical chemistry respectively from the University of Mysore in India and his Ph.D. in inorganic and radioanalytical chemistry from the Pennsylvania State University.

John Roecker is currently an active consultant in the Tank Waste Remediation Systems (TWRS) privatization program at Hanford, and has more than 39 years of experience in engineering, nuclear operations, and program management. This experience includes radioactive and hazardous waste management, nuclear chemical processing, space nuclear auxiliary power systems, breeder reactors, and commercial nuclear systems. Various companies at the Hanford site have employed Dr. Roecker since 1977. He served as manager of the TWRS Program Integration, deputy manager of defense waste remediation, and assistant to the vice president of environmental and waste management at the Westinghouse Hanford Company. Prior to working at Westinghouse Hanford, Mr. Roecker was employed by Rockwell International and Rockwell Hanford Operations where he served as director of Research and Engineering and as director of Waste Management Programs. He received a B.S. in engineering physics from the University of Illinois. He is a registered professional nuclear engineer and a member of the American Nuclear Society.

Ernest F. Ruppe is a retired vice president of petrochemicals of E.I. Du Pont de Nemours & Co., Inc. During his 39-year career with Du Pont, he served in a variety of engineering, research and development, manufacturing, and business management positions. In addition, he was Du Pont's first director of environmental affairs, was chairman of Du Pont International SA in Geneva, Switzerland, heading the company's European business, and later was responsible for the company's operation of the Savannah River site. He is a member of Columbia University's Engineering Council, and serves on several other nonprofit boards and committees. He received B.S. and M.S. degrees in mechanical engineering from Columbia.

Barry E. Scheetz, professor of materials, civil and nuclear engineering at the Pennsylvania State University, is active in research areas dealing with the chemistry of cementitious systems, including environmental remediation by the use of industrial by-products focusing on large-volume fly-ash-based cementitious grouts, nuclear and hazardous chemical waste management, crystal chemistry, and other areas of materials sciences. Among his many accomplishments, he received a national internship from the Argonne National Laboratory (1972), was a Visiting Scholar to China (1989), and won the Pennsylvania Governor's Environmental Award for Outstanding Achievement in Innovative Technology (1996). He is a member of the Mineralogical Society of America, the Materials Research Society, the American Ceramic Society, and Sigma Xi. Professor Scheetz has authored about 140 scientific publications, and holds 45 U.S. and foreign patents. He received a B.S. in chemical education from Bloomsburg State College (1967), a M.S. in geochemistry, and a Ph.D. in geochemistry and mineralogy from the Pennsylvania State University (1972 and 1976).

Anne Smith, a vice president at Charles River Associates (CRA), specializes in multidisciplinary policy assessment to help manage complex environmental and energy issues.

Dr. Smith's expertise is in analyzing costs, risks, and economic impacts of prominent air policy issues of the past decade, including regional haze, particulate matter, and ozone ambient standards, acid deposition, air toxics, global climate change, and emissions trading. She also has extensive experience in assessing risks and setting risk-based priorities for decisions associated with nuclear wastes, underground tanks, food product safety, and transportation. Her clients have included governments, research institutions, trade associations, multi-stakeholder organizations, and power companies. Dr. Smith played a prominent role in policy discussions related to the new PM2.5 and ozone NAAQS, providing influential analyses on their impacts to public health risks, costs, and regional economies. Her testimony was requested by the U.S. Congress in hearings on the new national ambient air quality standards and also on the proposed visibility (regional haze) standards. Prior to joining CRA, Dr. Smith was a vice president and manager of the policy practice at Decision Focus Incorporated. She also served as an economist in the Office of Policy Planning and Evaluation of the U.S. Environmental Protection Agency. Dr. Smith has a Ph.D. in economics from Stanford University, with a Ph.D. minor in engineering-economic systems.

Appendix E

Glossary

calcination: The decomposition of a material to a granular ceramic, usually involving an oxidizing atmosphere at an elevated temperature, for the purpose of changing its properties (e.g., to dehydrate or to form another phase). At the Idaho National Engineering and Environmental Laboratory, the calcination process injected waste into a fluidized bed to evaporate the water and decompose other material constituents into "calcine."

calcine: A general term for the granular ceramic generated by calcination. For the INEEL HLW program, it is a specific product from a fluidized bed calciner.

cladding: The outer protective layer, usually metal, over a fuel element.

contact-handled transuranic (CH-TRU): A classification of TRU waste with a surface contact surface dose rate <200 mrem/hour. The term derives from operational safety considerations; such waste could be handled safely, at least for short exposure times, by human contact at or near the surface of the waste container.

curie (Ci): A measure of radioactivity equal to 3.7×10^{10} disintegrations per second, a historical unit representing approximately the radioactivity of 1 gram of radium-226. The Systeme Internationale (SI) unit for radioactivity is the Becquerel, 1 disintegration per second.

Decontamination Factor (DF): The (dimensionless) ratio of the concentration of a species of interest that is in an original, input stream (to a process) to the concentration in a final, output stream. The typical quantity of interest that is used for this ratio is the radioactivity per unit volume or mass of carrier (e.g., solvent) material.

Environmental Assessment (EA) glass: The alkali borosilicate glass developed at the Savannah River Site, used in the environmental assessment of the Defense Waste Processing Facility (DWPF) there, and specified in the Waste Acceptance Product Specification (WAPS) as a benchmark for comparison, based on the PCT leaching test. The WAPS requires that HLW glasses be as durable (in PCT leach test analyses) as EA glass.

fluidized bed: A cushion of hot gas blown through the porous bottom of a container, used to float a powdered material as a means of drying, heating, quenching, or calcining [modified from Parker (1994)]. This process converts a material to a powder form, which behaves as if it were a fluid stirred by the action of high-velocity gases in a region in which a bed of large inert particles are similarly agitated. Both powder and gases exit the chamber and separate when the gas velocity lowers, enabling the powder to settle. Sometimes the conversion can be done without a particle bed by using the impact of the gas stream on the walls of the chamber or the self-comminution by the calcine itself.

high activity waste (HAW): Highly radioactive waste, used in this report to denote a concentrated fraction of the initial inventory of high-level waste.

high-level waste (HLW): High-level waste is (1) irradiated reactor fuel; (2) liquid wastes resulting from the operation of the first-cycle solvent extraction system, or equivalent, and the concentrated wastes from subsequent extraction cycles, or equivalent, in a facility for reprocessing irradiated reactor fuel; and (3) solids into which such liquid wastes have been converted (Source: 10 CFR 63 of the USNRC. in loose terms, HLW is the waste containing fission products and actinides that results from the reprocessing of spent nuclear fuel and which requires permanent isolation in a geologic repository.

highly enriched uranium (HEU): Uranium with more than 20 percent of the uranium-235 isotope.

ion exchange: A chemical reaction in which mobile hydrated ions of a solid are exchanged, equivalent for equivalent, for ions of like charge in solution [from Parker (1994)]. One or more of the ionic species in solution are selectively absorbed on a solid and retained.

low-level waste (LLW): Any radioactive waste that is not spent fuel or high-level waste.

mixed waste (MW): Waste that contains both chemically hazardous constituents regulated under RCRA and radioactive materials regulated under the Atomic Energy Act.

partitioning: A process that divides an input stream into two or more output streams. A mixture of solids can be partitioned based on differences in material properties. A liquid phase with two or more chemical species can be partitioned using selective media to segregate the species based on their differences in chemical reactivity.

raffinate: The aqueous solution remaining after the metal has been extracted by the solvent; the tailing of the solvent extraction system.

remote-handled transuranic (RH-TRU): A classification of TRU waste with a surface contact surface dose rate ≥ 200 mrem/hour. The term derives from operational safety considerations; such waste would need to be handled remotely, to provide adequate radiation shielding for workers.

reprocessing: Recovery of fissile and fertile material for further use from spent fuel by chemical separation of uranium and plutonium from other transuranic elements and fission products. Selected fission products may also be recovered. This operation results in the separation of wastes.

separations factor: The ratio of distribution coefficients (K_d) of two different elements in a given stage of a multi-stage solvent extraction process. A detailed, multi-stage process includes other effects such as the change in volume of solution due to other inputs such as scrub flows, that dilute the radionuclide species of interest irrespective of any separations achieved in extraction stages. These are usually expressed using an "extraction factor," which is the ratio, for two phases, of the product of each phase's distribution coefficient and flow rate.

solvent extraction: The separation of materials of different chemical types and solubilities by selective solvent action [from Parker (1994)]. This term is a general process for separating one or more chemicals solubilized in one solvent by the use of a second solvent that is (1) insoluble in the first solvent and (2) selective in its ability to bind one or more of the chemical constituents. After comixing and agitation of the two solvents, they separate, with the second having extracted the chemical species of interest.

spent nuclear fuel (SNF): Fuel that has been withdrawn from a nuclear reactor following irradiation, the constituent elements of which have not been separated by reprocessing.

sum of fractions: The USNRC's 10 CFR 61 lists values of allowed concentrations (i.e., activity in Ci/m^3) for a number of fission products and activation products for three different classes of low-level waste (A, B, C). For each individual radioisotope in a specific waste, the ratio of the actual activity to the respective limit represents the fraction of the allowed limit contributed by that isotope. Actual waste usually contains a mixture of several radioisotopes. To account for the contributions of all constituent radioisotopes in such mixtures, while remaining within the allowed limit for that waste class, the sum of these fractions for all radioisotopes present must be less than one. To allow margins for upsets, equipment failures, and other causes of process and feed variations, it is necessary in practice for the design point "sum of fractions" to be considerably less than 1 during normal operations. These considerations imply that the practical limit for each radioisotope is much less than the USNRC-stated limit.

transuranic (TRU) waste: Waste contaminated with transuranic elements with half-lives greater than 20 years, in concentrations greater than 100 nCi/g.

Appendix F

Acronyms and Symbols

A&E	architect and engineering, as in a company that performs design and construction services
AEA	Atomic Energy Act
ANN	aluminum nitrate
ANSI	American National Standards Institute
BSCP	Bin Set Closure Project
CERCLA	Comprehensive Environmental Response, Compensation, and Liability Act
CFR	Code of Federal Regulations
CH	contact handled
CMPO	octyl (phenyl)-N, N-di-isobutylcarbamoyl methyl phosphine oxide
CPS	cold pressing followed by sintering
CRTP	Calcine Retrieval and Transport Project
CSSF	Calcined Solids Storage Facility
DF	decontamination factor
DOE	U.S. Department of Energy
DOE-EM	DOE Office of Environmental Management
DWPF	Defense Waste Processing Facility
EA	Environmental Assessment
EIS	Environmental Impact Statement
EPA	U.S. Environmental Protection Agency
FUETAP	formed under elevated temperature and pressure
GCDF	Greater Confinement Disposal Facility
GOCO	government-owned, company-operated
HAW	high-activity waste
HEDPA	1-hydroxyethane 1,1-diphosphonic acid
HF	hydrofluoric
HIP	hot isostatic pressing
HLW	high-level waste
HUP	hot uniaxial pressing
ICPP	Idaho Chemical Processing Plant

INEEL	Idaho National Engineering and Environmental Laboratory
INTEC	Idaho Nuclear Technology and Engineering Center
K_d	distribution coefficient
LAW	low-activity waste
LDR	Land Disposal Restrictions (of RCRA)
LLW	low-level waste
LWR	light water reactor
MACT	maximal achievable controlled technology
MCC	Materials Characterization Center
MSRE	Molten Salt Reactor Experiment
MTHM	metric tons of heavy metal
NAE	National Academy of Engineering
NAS	National Academy of Sciences
NEPA	National Environmental Policy Act
NRC	National Research Council
NTS	Nevada Test Site
NWCF	New Waste Calcining Facility
NWPA	Nuclear Waste Policy Act
ORNL	Oak Ridge National Laboratory
PCT	product consistency test
PUREX	plutonium and uranium recovery by extraction
RCRA	Resource Conservation and Recovery Act
RFP	request for proposals
RH	remote handled
SBW	sodium-bearing waste
SEM	scanning electron microscopy
SLS	solid–liquid separation
SNF	spent nuclear fuel
SREX	strontium extraction
SSF	simulated spent fuel
TBP	tributyl phosphate
TCLP	toxic characteristic leaching procedure
TRU	transuranic
TRUEX	transuranium extraction
UDS	undissolved solids
USNRC	U.S. Nuclear Regulatory Commission
WAPS	waste acceptance product specification
WCF	Waste Calcining Facility
WIPP	Waste Isolation Pilot Plant
XRD	x-ray diffraction

Appendix G

Portions of the 1995 Settlement Agreement

The 1995 Settlement Agreement (Lodge, 1995) covered many activities at the INEEL for inventories of DOE and Navy spent fuel, DOE transuranic waste, DOE high-level waste, and other treatable mixed waste. The quotations shown here are those parts of the agreement most directly relevant to the DOE HLW program to remediate the calcine and SBW.

"DOE shall treat all high-level waste currently at INEL so that it is ready to be moved out of Idaho for disposal by a target date of 2035."

And

"DOE agrees to treat spent fuel, high-level waste, and transuranic wastes in Idaho requiring treatment so as to permit ultimate disposal outside the State of Idaho."

And

"DOE shall commence calcination of sodium-bearing liquid high-level wastes by June 1, 2001. DOE shall complete calcination of sodium-bearing liquid high-level wastes by December 31, 2012."

And

"DOE shall accelerate efforts to evaluate alternatives for the treatment of calcined waste so as to put it into a form suitable for transport to a permanent repository or interim storage facility outside Idaho. To support this effort, DOE shall solicit proposals for feasibility studies by July 1, 1997. By December 31, 1999, DOE shall commence negotiating a plan and schedule with the State of Idaho for calcined waste treatment. The plan and schedule shall provide for completion of the treatment of all calcined waste located at INEL by a date established by the Record of Decision for the Environmental Impact Statement that analyzes the alternatives for treatment of such waste. Such Record of Decision shall be issued not later than December 31, 2009. It is presently contemplated by DOE that the plan and schedule shall provide for the completion of the treatment of all calcined waste located at INEL by a target date of December 31, 2035. The State expressly reserves its right to seek appropriate relief from the Court in the event that the date established in the Record of Decision for the Environmental Impact Statement that analyzes the alternatives for treatment of such waste is significantly later than DOE's target date. In support of the effort to treat such waste, DOE shall submit to the State of Idaho its application for a RCRA (or statutory equivalent) Part B permit by December 1, 2012."

And

"In the event any required NEPA analysis results in the selection after October 16, 1995, of an action which conflicts with any action identified in this Agreement, DOE or the Navy may request a modification of this Agreement to conform the action in the Agreement to that selected action. Approval of such modification shall not be unreasonably withheld. If the State refuses to accept the requested modification, DOE or the Navy may seek relief from the Court. On motion of any party, the Court may extend the time for DOE or the Navy to perform until the Court has decided whether to grant relief. If the Court determines that the State has unreasonably withheld approval, the Agreement shall be conformed to the selected action. If the Court determines that the State has reasonably withheld approval, the time for DOE or the Navy to perform the action at issue shall be as set forth in this Agreement and subject to enforcement as set forth in Section K.1."